The Elements of Style

THE
ELEMENTS
OF
Style

BY
WILLIAM STRUNK Jr.

With Revisions, an Introduction,
and a Chapter on Writing

BY
E. B. WHITE

FOURTH EDITION

PEARSON

Boston Columbus Indianapolis New York San Francisco Upper Saddle River
Amsterdam Cape Town Dubai London Madrid Milan Munich Paris Montreal Toronto
Delhi Mexico City São Paulo Sydney Hong Kong Seoul Singapore Taipei Tokyo

Library of Congress Cataloging-in-Publication Data

Strunk, William, 1869–1946.
  The elements of style / by William Strunk, Jr. ; with revisions,
an introduction, and a chapter on writing by E. B. White. — 4th ed.
    p.   cm.
  Includes index.
  ISBN 0-205-30902-X (paperback). — ISBN 0-205-31342-6 (casebound)
  1. English language—Rhetoric.   2. English language—Style.
3. Report writing.   I. White, E. B. (Elwyn Brooks), 1899–   .
II. Title.
PE1408.S772   1999
808'.042—dc21   99-16419
                                                             CIP

52 2021

CV 03.08.2021 0845

Contents

Foreword

by Roger Angell

THE FIRST writer I watched at work was my stepfather, E. B. White. Each Tuesday morning, he would close his study door and sit down to write the "Notes and Comment" page for *The New Yorker*. The task was familiar to him—he was required to file a few hundred words of editorial or personal commentary on some topic in or out of the news that week—but the sounds of his typewriter from his room came in hesitant bursts, with long silences in between. Hours went by. Summoned at last for lunch, he was silent and preoccupied, and soon excused himself to get back to the job. When the copy went off at last, in the afternoon RFD pouch—we were in Maine, a day's mail away from New York—he rarely seemed satisfied. "It isn't good enough," he said sometimes. "I wish it were better."

Writing is hard, even for authors who do it all the time. Less frequent practitioners—the job applicant; the business executive with an annual report to get out; the high school senior with a Faulkner assignment; the graduate-school student with her thesis proposal; the writer of a letter of condolence—often get stuck in an awkward passage or find a muddle on their screens, and then blame themselves. What should be easy and flowing looks tangled or feeble or overblown—not what was meant at all. What's wrong with me, each one thinks. Why can't I get this right?

It was this recurring question, put to himself, that must have inspired White to revive and add to a textbook by an English professor of his, Will Strunk Jr., that he had first read in college, and to get it published. The result, this quiet book, has been in print for forty years, and has offered more than ten million writers a helping hand. White knew that a compendium of specific tips—about singular and plural verbs, parentheses, the "that"–"which" scuffle, and many others—could clear up a recalcitrant sentence or subclause when quickly reconsulted, and that the larger principles needed to be kept in plain sight, like a wall sampler.

How simple they look, set down here in White's last chapter: "Write in a way that comes naturally," "Revise and rewrite," "Do not explain too much," and the rest; above all, the cleansing, clarion "Be clear." How often I have turned to them, in the book or in my mind, while trying to start or unblock or revise some piece of my own writing! They help—they really do. They work. They are the way.

E. B. White's prose is celebrated for its ease and clarity— just think of *Charlotte's Web*—but maintaining this standard required endless attention. When the new issue of *The New Yorker* turned up in Maine, I sometimes saw him reading his "Comment" piece over to himself, with only a slightly different expression than the one he'd worn on the day it went off. Well, O.K., he seemed to be saying. At least I got the elements right.

This edition has been modestly updated, with word processors and air conditioners making their first appearance among White's references, and with a light redistribution of genders to permit a feminine pronoun or female farmer to take their places among the males who once innocently served him. Sylvia Plath has knocked Keats out of the box, and I notice that "America" has become "this country" in a sample text, to forestall a subsequent and possibly demeaning "she" in the same paragraph. What is not here is anything about E-mail—the rules-free, lower-case flow that cheerfully keeps us in touch these days. E-mail is conversation, and it may be replacing the sweet and endless talking we once sustained (and tucked away) within the informal letter. But we are all writers and readers as well as communicators, with the need at times to please and satisfy ourselves (as White put it) with the clear and almost perfect thought.

Introduction*

AT THE close of the first World War, when I was a student at Cornell, I took a course called English 8. My professor was William Strunk Jr. A textbook required for the course was a slim volume called *The Elements of Style,* whose author was the professor himself. The year was 1919. The book was known on the campus in those days as "the little book," with the stress on the word "little." It had been privately printed by the author.

I passed the course, graduated from the university, and forgot the book but not the professor. Some thirty-eight years later, the book bobbed up again in my life when Macmillan commissioned me to revise it for the college market and the general trade. Meantime, Professor Strunk had died.

The Elements of Style, when I reexamined it in 1957, seemed to me to contain rich deposits of gold. It was Will Strunk's *parvum opus,* his attempt to cut the vast tangle of English rhetoric down to size and write its rules and principles on the head of a pin. Will himself had hung the tag "little" on the book; he referred to it sardonically and with secret pride as "the *little* book," always giving the word "little" a special twist, as though he were putting a spin on a ball. In its original form, it was a forty-three page summation of the case for cleanliness, accuracy, and brevity in the use of English. Today, fifty-two years later, its vigor is

*E. B. White wrote this introduction for the 1979 edition.

xiii

unimpaired, and for sheer pith I think it probably sets a record that is not likely to be broken. Even after I got through tampering with it, it was still a tiny thing, a barely tarnished gem. Seven rules of usage, eleven principles of composition, a few matters of form, and a list of words and expressions commonly misused—that was the sum and substance of Professor Strunk's work. Somewhat audaciously, and in an attempt to give my publisher his money's worth, I added a chapter called "An Approach to Style," setting forth my own prejudices, my notions of error, my articles of faith. This chapter (Chapter V) is addressed particularly to those who feel that English prose composition is not only a necessary skill but a sensible pursuit as well—a way to spend one's days. I think Professor Strunk would not object to that.

A second edition of the book was published in 1972. I have now completed a third revision. Chapter IV has been refurbished with words and expressions of a recent vintage; four rules of usage have been added to Chapter I. Fresh examples have been added to some of the rules and principles, amplification has reared its head in a few places in the text where I felt an assault could successfully be made on the bastions of its brevity, and in general the book has received a thorough overhaul—to correct errors, delete bewhiskered entries, and enliven the argument.

Professor Strunk was a positive man. His book contains rules of grammar phrased as direct orders. In the main I have not tried to soften his commands, or modify his pronouncements, or remove the special objects of his scorn. I have tried, instead, to preserve the flavor of his discontent while slightly enlarging the scope of the discussion. *The Elements of Style* does not pretend to survey the whole field. Rather it proposes to give in brief space the principal requirements of plain English style. It concentrates on fundamentals: the rules of usage and principles of composition most commonly violated.

The reader will soon discover that these rules and principles are in the form of sharp commands, Sergeant Strunk snapping orders to his platoon. "Do not join independent clauses with a comma." (Rule 5.) "Do not break sentences in two." (Rule 6.) "Use the active voice." (Rule 14.) "Omit needless words." (Rule 17.) "Avoid a succession of loose sentences." (Rule 18.) "In summaries, keep to one tense." (Rule 21.) Each rule or principle is followed by a short hortatory essay, and usually the exhortation is followed by, or interlarded with, examples in parallel columns—the true vs. the false, the right vs. the wrong, the timid vs. the bold, the ragged vs. the trim. From every line there peers out at me the puckish face of my professor, his short hair parted neatly in the middle and combed down over his forehead, his eyes blinking incessantly behind steel-rimmed spectacles as though he had just emerged into strong light, his lips nibbling each other like nervous horses, his smile shuttling to and fro under a carefully edged mustache.

"Omit needless words!" cries the author on page 23, and into that imperative Will Strunk really put his heart and soul. In the days when I was sitting in his class, he omitted so many needless words, and omitted them so forcibly and with such eagerness and obvious relish, that he often seemed in the position of having shortchanged himself—a man left with nothing more to say yet with time to fill, a radio prophet who had out-distanced the clock. Will Strunk got out of this predicament by a simple trick: he uttered every sentence three times. When he delivered his oration on brevity to the class, he leaned forward over his desk, grasped his coat lapels in his hands, and, in a husky, conspiratorial voice, said, "Rule Seventeen. Omit needless words! Omit needless words! Omit needless words!"

He was a memorable man, friendly and funny. Under the remembered sting of his kindly lash, I have been trying to omit needless words since 1919, and although there are still many words that cry for omission and the huge task will never be accomplished, it is exciting to me to reread the masterly Strunkian elaboration of this noble theme. It goes:

> Vigorous writing is concise. A sentence should contain no unnecessary words, a paragraph no unnecessary sentences, for the same reason that a drawing should have no

unnecessary lines and a machine no unnecessary parts.
This requires not that the writer make all sentences short
or avoid all detail and treat subjects only in outline, but
that every word tell.

There you have a short, valuable essay on the nature and
beauty of brevity—fifty-nine words that could change the
world. Having recovered from his adventure in prolixity
(fifty-nine words were a lot of words in the tight world of
William Strunk Jr.), the professor proceeds to give a few
quick lessons in pruning. Students learn to cut the dead-
wood from "this is a subject that," reducing it to "this subject," a saving of three words. They learn to trim "used for
fuel purposes" down to "used for fuel." They learn that they
are being chatterboxes when they say "the question as to
whether" and that they should just say "whether"—a saving
of four words out of a possible five.

The professor devotes a special paragraph to the vile
expression *the fact that,* a phrase that causes him to quiver
with revulsion. The expression, he says, should be "revised
out of every sentence in which it occurs." But a shadow of
gloom seems to hang over the page, and you feel that he
knows how hopeless his cause is. I suppose I have written
the fact that a thousand times in the heat of composition,
revised it out maybe five hundred times in the cool after-
math. To be batting only .500 this late in the season, to fail
half the time to connect with this fat pitch, saddens me, for
it seems a betrayal of the man who showed me how to swing
at it and made the swinging seem worthwhile.

I treasure *The Elements of Style* for its sharp advice, but
I treasure it even more for the audacity and self-confidence
of its author. Will knew where he stood. He was so sure of
where he stood, and made his position so clear and so plau-
sible, that his peculiar stance has continued to invigorate
me—and, I am sure, thousands of other ex-students—during
the years that have intervened since our first encounter. He
had a number of likes and dislikes that were almost as
whimsical as the choice of a necktie, yet he made them
seem utterly convincing. He disliked the word *forceful* and

advised us to use *forcible* instead. He felt that the word *clever* was greatly overused: "It is best restricted to ingenuity displayed in small matters." He despised the expression *student body,* which he termed gruesome, and made a special trip downtown to the *Alumni News* office one day to protest the expression and suggest that *studentry* be substituted—a coinage of his own, which he felt was similar to *citizenry.* I am told that the *News* editor was so charmed by the visit, if not by the word, that he ordered the student body buried, never to rise again. *Studentry* has taken its place. It's not much of an improvement, but it does sound less cadaverous, and it made Will Strunk quite happy.

Some years ago, when the heir to the throne of England was a child, I noticed a headline in the *Times* about Bonnie Prince Charlie: "CHARLES' TONSILS OUT." Immediately Rule 1 leapt to mind.

1. Form the possessive singular of nouns by adding *'s.* Follow this rule whatever the final consonant. Thus write,

Charles's friend

Burns's poems

the witch's malice

Clearly, Will Strunk had foreseen, as far back as 1918, the dangerous tonsillectomy of a prince, in which the surgeon removes the tonsils and the *Times* copy desk removes the final *s.* He started his book with it. I commend Rule 1 to the *Times,* and I trust that Charles's throat, not Charles' throat, is in fine shape today.

Style rules of this sort are, of course, somewhat a matter of individual preference, and even the established rules of grammar are open to challenge. Professor Strunk, although one of the most inflexible and choosy of men, was quick to acknowledge the fallacy of inflexibility and the danger of doctrine. "It is an old observation," he wrote, "that the best writers sometimes disregard the rules of rhetoric. When they do so, however, the reader will usually find in the sentence some compensating merit, attained at the cost of the violation. Unless he is certain of doing as well, he will probably do best to follow the rules."

It is encouraging to see how perfectly a book, even a dusty rule book, perpetuates and extends the spirit of a man. Will Strunk loved the clear, the brief, the bold, and his book is clear, brief, bold. Boldness is perhaps its chief distinguishing mark. On page 26, explaining one of his parallels, he says, "The lefthand version gives the impression that the writer is undecided or timid, apparently unable or afraid to choose one form of expression and hold to it." And his original Rule 11 was "Make definite assertions." That was Will all over. He scorned the vague, the tame, the colorless, the irresolute. He felt it was worse to be irresolute than to be wrong. I remember a day in class when he leaned far forward, in his characteristic pose—the pose of a man about to impart a secret—and croaked, "If you don't know how to pronounce a word, say it loud! If you don't know how to pronounce a word, say it loud!" This comical piece of advice struck me as sound at the time, and I still respect it. Why compound ignorance with inaudibility? Why run and hide?

All through *The Elements of Style* one finds evidences of the author's deep sympathy for the reader. Will felt that the reader was in serious trouble most of the time, floundering in a swamp, and that it was the duty of anyone attempting to write English to drain this swamp quickly and get the reader up on dry ground, or at least to throw a rope. In revising the text, I have tried to hold steadily in mind this belief of his, this concern for the bewildered reader.

In the English classes of today, "the little book" is surrounded by longer, lower textbooks—books with permissive steering and automatic transitions. Perhaps the book has become something of a curiosity. To me, it still seems to maintain its original poise, standing, in a drafty time, erect, resolute, and assured. I still find the Strunkian wisdom a comfort, the Strunkian humor a delight, and the Strunkian attitude toward right-and-wrong a blessing undisguised.

E. B. WHITE

The Elements of Style

I

Elementary Rules of Usage

1. ***Form the possessive singular of nouns by adding 's.***

Follow this rule whatever the final consonant. Thus write,

> Charles's friend
>
> Burns's poems
>
> the witch's malice

Exceptions are the possessives of ancient proper names ending in *-es* and *-is*, the possessive *Jesus'*, and such forms as *for conscience' sake, for righteousness' sake*. But such forms as *Moses' Laws, Isis' temple* are commonly replaced by

> the laws of Moses
>
> the temple of Isis

The pronominal possessives *hers, its, theirs, yours,* and *ours* have no apostrophe. Indefinite pronouns, however, use the apostrophe to show possession.

> one's rights
>
> somebody else's umbrella

A common error is to write *it's* for *its*, or vice versa. The first is a contraction, meaning "it is." The second is a possessive.

> It's a wise dog that scratches its own fleas.

1

2. *In a series of three or more terms with a single conjunction, use a comma after each term except the last.*

Thus write,

> red, white, and blue
>
> gold, silver, or copper
>
> He opened the letter, read it, and made a note of its contents.

This comma is often referred to as the "serial" comma. In the names of business firms the last comma is usually omitted. Follow the usage of the individual firm.

> Little, Brown and Company
>
> Donaldson, Lufkin & Jenrette

3. *Enclose parenthetic expressions between commas.*

> The best way to see a country, unless you are pressed for time, is to travel on foot.

This rule is difficult to apply; it is frequently hard to decide whether a single word, such as *however,* or a brief phrase is or is not parenthetic. If the interruption to the flow of the sentence is but slight, the commas may be safely omitted. But whether the interruption is slight or considerable, never omit one comma and leave the other. There is no defense for such punctuation as

> Marjorie's husband, Colonel Nelson paid us a visit yesterday.

or

> My brother you will be pleased to hear, is now in perfect health.

Dates usually contain parenthetic words or figures. Punctuate as follows:

> February to July, 1992

April 6, 1986

Wednesday, November 14, 1990

Note that it is customary to omit the comma in

6 April 1988

The last form is an excellent way to write a date; the figures are separated by a word and are, for that reason, quickly grasped.

A name or a title in direct address is parenthetic.

If, Sir, you refuse, I cannot predict what will happen.

Well, Susan, this is a fine mess you are in.

The abbreviations *etc., i.e.,* and *e.g.,* the abbreviations for academic degrees, and titles that follow a name are parenthetic and should be punctuated accordingly.

Letters, packages, etc., should go here.

Horace Fulsome, Ph.D., presided.

Rachel Simonds, Attorney

The Reverend Harry Lang, S.J.

No comma, however, should separate a noun from a restrictive term of identification.

Billy the Kid

The novelist Jane Austen

William the Conqueror

The poet Sappho

Although *Junior,* with its abbreviation *Jr.,* has commonly been regarded as parenthetic, logic suggests that it is, in fact, restrictive and therefore not in need of a comma.

James Wright Jr.

Nonrestrictive relative clauses are parenthetic, as are similar clauses introduced by conjunctions indicating time or place. Commas are therefore needed. A nonrestrictive clause is one that does not serve to identify or define the antecedent noun.

> The audience, which had at first been indifferent, became more and more interested.

> In 1769, when Napoleon was born, Corsica had but recently been acquired by France.

> Nether Stowey, where Coleridge wrote *The Rime of the Ancient Mariner,* is a few miles from Bridgewater.

In these sentences, the clauses introduced by *which, when,* and *where* are nonrestrictive; they do not limit or define, they merely add something. In the first example, the clause introduced by *which* does not serve to tell which of several possible audiences is meant; the reader presumably knows that already. The clause adds, parenthetically, a statement supplementing that in the main clause. Each of the three sentences is a combination of two statements that might have been made independently.

> The audience was at first indifferent. Later it became more and more interested.

> Napoleon was born in 1769. At that time Corsica had but recently been acquired by France.

> Coleridge wrote *The Rime of the Ancient Mariner* at Nether Stowey. Nether Stowey is a few miles from Bridgewater.

Restrictive clauses, by contrast, are not parenthetic and are not set off by commas. Thus,

> People who live in glass houses shouldn't throw stones.

Here the clause introduced by *who* does serve to tell which people are meant; the sentence, unlike the sentences above, cannot be split into two independent statements. The same principle of comma use applies to participial phrases and to appositives.

> People sitting in the rear couldn't hear. (*restrictive*)

> Uncle Bert, being slightly deaf, moved forward. (*nonrestrictive*)

> My cousin Bob is a talented harpist. (*restrictive*)

> Our oldest daughter, Mary, sings. (*nonrestrictive*)

When the main clause of a sentence is preceded by a phrase or a subordinate clause, use a comma to set off these elements.

> Partly by hard fighting, partly by diplomatic skill, they enlarged their dominions to the east and rose to royal rank with the possession of Sicily.

4. *Place a comma before a conjunction introducing an independent clause.*

> The early records of the city have disappeared, and the story of its first years can no longer be reconstructed.

> The situation is perilous, but there is still one chance of escape.

Two-part sentences of which the second member is introduced by *as* (in the sense of "because"), *for, or, nor,* or *while* (in the sense of "and at the same time") likewise require a comma before the conjunction.

If a dependent clause, or an introductory phrase requiring to be set off by a comma, precedes the second independent clause, no comma is needed after the conjunction.

> The situation is perilous, but if we are prepared to act promptly, there is still one chance of escape.

When the subject is the same for both clauses and is expressed only once, a comma is useful if the connective is *but.* When the connective is *and,* the comma should be omitted if the relation between the two statements is close or immediate.

> I have heard the arguments, but am still unconvinced.

> He has had several years' experience and is thoroughly competent.

5. *Do not join independent clauses with a comma.*

If two or more clauses grammatically complete and not joined by a conjunction are to form a single compound sentence, the proper mark of punctuation is a semicolon.

> Mary Shelley's works are entertaining; they are full of engaging ideas.

> It is nearly half past five; we cannot reach town before dark.

It is, of course, equally correct to write each of these as two sentences, replacing the semicolons with periods.

> Mary Shelley's works are entertaining. They are full of engaging ideas.

> It is nearly half past five. We cannot reach town before dark.

If a conjunction is inserted, the proper mark is a comma. (Rule 4.)

> Mary Shelley's works are entertaining, for they are full of engaging ideas.

> It is nearly half past five, and we cannot reach town before dark.

A comparison of the three forms given above will show clearly the advantage of the first. It is, at least in the examples given, better than the second form because it suggests the close relationship between the two statements in a way that the second does not attempt, and better than the third because it is briefer and therefore more forcible. Indeed, this simple method of indicating relationship between statements is one of the most useful devices of composition. The relationship, as above, is commonly one of cause and consequence.

Note that if the second clause is preceded by an adverb, such as *accordingly, besides, then, therefore,* or *thus,* and not by a conjunction, the semicolon is still required.

> I had never been in the place before; besides, it was dark as a tomb.

An exception to the semicolon rule is worth noting here. A comma is preferable when the clauses are very short and alike in form, or when the tone of the sentence is easy and conversational.

> Man proposes, God disposes.

> The gates swung apart, the bridge fell, the portcullis was drawn up.

> I hardly knew him, he was so changed.

> Here today, gone tomorrow.

6. *Do not break sentences in two.*

In other words, do not use periods for commas.

> I met them on a Cunard liner many years ago. Coming home from Liverpool to New York.

> She was an interesting talker. A woman who had traveled all over the world and lived in half a dozen countries.

In both these examples, the first period should be replaced by a comma and the following word begun with a small letter.

It is permissible to make an emphatic word or expression serve the purpose of a sentence and to punctuate it accordingly:

> Again and again he called out. No reply.

The writer must, however, be certain that the emphasis is warranted, lest a clipped sentence seem merely a blunder in syntax or in punctuation. Generally speaking, the place for broken sentences is in dialogue, when a character happens to speak in a clipped or fragmentary way.

Rules 3, 4, 5, and 6 cover the most important principles that govern punctuation. They should be so thoroughly mastered that their application becomes second nature.

7. *Use a colon after an independent clause to introduce a list of particulars, an appositive, an amplification, or an illustrative quotation.*

A colon tells the reader that what follows is closely related to the preceding clause. The colon has more effect than the comma, less power to separate than the semicolon, and more formality than the dash. It usually follows an independent clause and should not separate a verb from its complement or a preposition from its object. The examples in the lefthand column, below, are wrong; they should be rewritten as in the righthand column.

Your dedicated whittler requires: a knife, a piece of wood, and a back porch.	Your dedicated whittler requires three props: a knife, a piece of wood, and a back porch.
Understanding is that penetrating quality of knowledge that grows from: theory, practice, conviction, assertion, error, and humiliation.	Understanding is that penetrating quality of knowledge that grows from theory, practice, conviction, assertion, error, and humiliation.

Join two independent clauses with a colon if the second interprets or amplifies the first.

> But even so, there was a directness and dispatch about animal burial: there was no stopover in the undertaker's foul parlor, no wreath or spray.

A colon may introduce a quotation that supports or contributes to the preceding clause.

> The squalor of the streets reminded her of a line from Oscar Wilde: "We are all in the gutter, but some of us are looking at the stars."

The colon also has certain functions of form: to follow the salutation of a formal letter, to separate hour from minute in a notation of time, and to separate the title of a work from its subtitle or a Bible chapter from a verse.

> Dear Mr. Montague:
>
> departs at 10:48 P.M.
>
> *Practical Calligraphy: An Introduction to Italic Script*
>
> Nehemiah 11:7

8. *Use a dash to set off an abrupt break or interruption and to announce a long appositive or summary.*

A dash is a mark of separation stronger than a comma, less formal than a colon, and more relaxed than parentheses.

> His first thought on getting out of bed—if he had any thought at all—was to get back in again.

> The rear axle began to make a noise—a grinding, chattering, teeth-gritting rasp.

> The increasing reluctance of the sun to rise, the extra nip in the breeze, the patter of shed leaves dropping—all the evidences of fall drifting into winter were clearer each day.

Use a dash only when a more common mark of punctuation seems inadequate.

Her father's suspicions proved well-founded—it was not Edward she cared for—it was San Francisco.	Her father's suspicions proved well-founded. It was not Edward she cared for, it was San Francisco.
Violence—the kind you see on television—is not honestly violent—there lies its harm.	Violence, the kind you see on television, is not honestly violent. There lies its harm.

9. *The number of the subject determines the number of the verb.*

Words that intervene between subject and verb do not affect the number of the verb.

The bittersweet flavor of youth—its trials, its joys, its adventures, its challenges— are not soon forgotten.	The bittersweet flavor of youth—its trials, its joys, its adventures, its challenges —is not soon forgotten.

A common blunder is the use of a singular verb form in a relative clause following "one of . . ." or a similar expression when the relative is the subject.

One of the ablest scientists who has attacked this problem	One of the ablest scientists who have attacked this problem
One of those people who is never ready on time	One of those people who are never ready on time

Use a singular verb form after *each, either, everyone, everybody, neither, nobody, someone.*

Everybody thinks he has a unique sense of humor.

Although both clocks strike cheerfully, neither keeps good time.

With *none,* use the singular verb when the word means "no one" or "not one."

None of us are perfect.	None of us is perfect.

A plural verb is commonly used when *none* suggests more than one thing or person.

None are so fallible as those who are sure they're right.

A compound subject formed of two or more nouns joined by *and* almost always requires a plural verb.

The walrus and the carpenter were walking close at hand.

But certain compounds, often clichés, are so inseparable they are considered a unit and so take a singular verb, as do compound subjects qualified by *each* or *every.*

The long and the short of it is . . .

Bread and butter was all she served.

Give and take is essential to a happy household.

Every window, picture, and mirror was smashed.

A singular subject remains singular even if other nouns are connected to it by *with, as well as, in addition to, except, together with,* and *no less than.*

His speech as well as his manner is objectionable.

A linking verb agrees with the number of its subject.

> What is wanted is a few more pairs of hands.
>
> The trouble with truth is its many varieties.

Some nouns that appear to be plural are usually construed as singular and given a singular verb.

> Politics is an art, not a science.
>
> The Republican Headquarters is on this side of the tracks.

But

> The general's quarters are across the river.

In these cases the writer must simply learn the idioms. The contents of a book is singular. The contents of a jar may be either singular or plural, depending on what's in the jar— jam or marbles.

10. *Use the proper case of pronoun.*

The personal pronouns, as well as the pronoun *who,* change form as they function as subject or object.

> Will Jane or he be hired, do you think?
>
> The culprit, it turned out, was he.
>
> We heavy eaters would rather walk than ride.
>
> Who knocks?
>
> Give this work to whoever looks idle.

In the last example, *whoever* is the subject of *looks idle;* the object of the preposition *to* is the entire clause *whoever looks idle.* When *who* introduces a subordinate clause, its case depends on its function in that clause.

Virgil Soames is the candidate whom we think will win.	Virgil Soames is the candidate who we think will win. [We think *he* will win.]
Virgil Soames is the candidate who we hope to elect.	Virgil Soames is the candidate whom we hope to elect. [We hope to elect *him.*]

A pronoun in a comparison is nominative if it is the subject of a stated or understood verb.

> Sandy writes better than I. (Than I write.)

In general, avoid "understood" verbs by supplying them.

| I think Horace admires Jessica more than I. | I think Horace admires Jessica more than I do. |
| Polly loves cake more than me. | Polly loves cake more than she loves me. |

The objective case is correct in the following examples.

> The ranger offered Shirley and him some advice on campsites.
>
> They came to meet the Baldwins and us.
>
> Let's talk it over between us, then, you and me.
>
> Whom should I ask?
>
> A group of us taxpayers protested.

Us in the last example is in apposition to taxpayers, the object of the preposition *of.* The wording, although grammatically defensible, is rarely apt. "A group of us protested as taxpayers" is better, if not exactly equivalent.

Use the simple personal pronoun as a subject.

| Blake and myself stayed home. | Blake and I stayed home. |
| Howard and yourself brought the lunch, I thought. | Howard and you brought the lunch, I thought. |

The possessive case of pronouns is used to show ownership. It has two forms: the adjectival modifier, *your* hat, and the noun form, a hat of *yours.*

> The dog has buried one of your gloves and one of mine in the flower bed.

Gerunds usually require the possessive case.

> Mother objected to our driving on the icy roads.

A present participle as a verbal, on the other hand, takes the objective case.

> They heard him singing in the shower.

The difference between a verbal participle and a gerund is not always obvious, but note what is really said in each of the following.

> Do you mind me asking a question?

> Do you mind my asking a question?

In the first sentence, the queried objection is to *me,* as opposed to other members of the group, asking a question. In the second example, the issue is whether a question may be asked at all.

11. *A participial phrase at the beginning of a sentence must refer to the grammatical subject.*

> Walking slowly down the road, he saw a woman accompanied by two children.

The word *walking* refers to the subject of the sentence, not to the woman. To make it refer to the woman, the writer must recast the sentence.

> He saw a woman, accompanied by two children, walking slowly down the road.

Participial phrases preceded by a conjunction or by a preposition, nouns in apposition, adjectives, and adjective phrases come under the same rule if they begin the sentence.

On arriving in Chicago, his friends met him at the station.	On arriving in Chicago, he was met at the station by his friends.
A soldier of proved valor, they entrusted him with the defense of the city.	A soldier of proved valor, he was entrusted with the defense of the city.
Young and inexperienced, the task seemed easy to me.	Young and inexperienced, I thought the task easy.

| Without a friend to counsel him, the temptation proved irresistible. | Without a friend to counsel him, he found the temptation irresistible. |

Sentences violating Rule 11 are often ludicrous:

Being in a dilapidated condition, I was able to buy the house very cheap.

Wondering irresolutely what to do next, the clock struck twelve.

II

Elementary Principles of Composition

12. *Choose a suitable design and hold to it.*

A basic structural design underlies every kind of writing. Writers will in part follow this design, in part deviate from it, according to their skills, their needs, and the unexpected events that accompany the act of composition. Writing, to be effective, must follow closely the thoughts of the writer, but not necessarily in the order in which those thoughts occur. This calls for a scheme of procedure. In some cases, the best design is no design, as with a love letter, which is simply an outpouring, or with a casual essay, which is a ramble. But in most cases, planning must be a deliberate prelude to writing. The first principle of composition, therefore, is to foresee or determine the shape of what is to come and pursue that shape.

A sonnet is built on a fourteen-line frame, each line containing five feet. Hence, sonneteers know exactly where they are headed, although they may not know how to get there. Most forms of composition are less clearly defined, more flexible, but all have skeletons to which the writer will bring the flesh and the blood. The more clearly the writer perceives the shape, the better are the chances of success.

13. *Make the paragraph the unit of composition.*

The paragraph is a convenient unit; it serves all forms of literary work. As long as it holds together, a paragraph may be of any length—a single, short sentence or a passage of great duration.

If the subject on which you are writing is of slight extent, or if you intend to treat it briefly, there may be no need to divide it into topics. Thus, a brief description, a brief book review, a brief account of a single incident, a narrative merely outlining an action, the setting forth of a single idea—any one of these is best written in a single paragraph. After the paragraph has been written, examine it to see whether division will improve it.

Ordinarily, however, a subject requires division into topics, each of which should be dealt with in a paragraph. The object of treating each topic in a paragraph by itself is, of course, to aid the reader. The beginning of each paragraph is a signal that a new step in the development of the subject has been reached.

As a rule, single sentences should not be written or printed as paragraphs. An exception may be made of sentences of transition, indicating the relation between the parts of an exposition or argument.

In dialogue, each speech, even if only a single word, is usually a paragraph by itself; that is, a new paragraph begins with each change of speaker. The application of this rule when dialogue and narrative are combined is best learned from examples in well-edited works of fiction. Sometimes a writer, seeking to create an effect of rapid talk or for some other reason, will elect not to set off each speech in a separate paragraph and instead will run speeches together. The common practice, however, and the one that serves best in most instances, is to give each speech a paragraph of its own.

As a rule, begin each paragraph either with a sentence that suggests the topic or with a sentence that helps the transition. If a paragraph forms part of a larger composition, its relation to what precedes, or its function as a part of the whole, may need to be expressed. This can sometimes be done by a mere word or phrase (*again, therefore, for the same reason*) in the first sentence. Sometimes, however, it is expedient to get into the topic slowly, by way of a sentence or two of introduction or transition.

In narration and description, the paragraph sometimes begins with a concise, comprehensive statement serving to hold together the details that follow.

The breeze served us admirably.

The campaign opened with a series of reverses.

The next ten or twelve pages were filled with a curious set of entries.

But when this device, or any device, is too often used, it becomes a mannerism. More commonly, the opening sentence simply indicates by its subject the direction the paragraph is to take.

At length I thought I might return toward the stockade.

He picked up the heavy lamp from the table and began to explore.

Another flight of steps, and they emerged on the roof.

In animated narrative, the paragraphs are likely to be short and without any semblance of a topic sentence, the writer rushing headlong, event following event in rapid succession. The break between such paragraphs merely serves the purpose of a rhetorical pause, throwing into prominence some detail of the action.

In general, remember that paragraphing calls for a good eye as well as a logical mind. Enormous blocks of print look formidable to readers, who are often reluctant to tackle them. Therefore, breaking long paragraphs in two, even if it is not necessary to do so for sense, meaning, or logical development, is often a visual help. But remember, too, that firing off many short paragraphs in quick succession can be distracting. Paragraph breaks used only for show read like the writing of commerce or of display advertising. Moderation and a sense of order should be the main considerations in paragraphing.

14. *Use the active voice.*

The active voice is usually more direct and vigorous than the passive:

> I shall always remember my first visit to Boston.

This is much better than

> My first visit to Boston will always be remembered by me.

The latter sentence is less direct, less bold, and less concise. If the writer tries to make it more concise by omitting "by me,"

> My first visit to Boston will always be remembered,

it becomes indefinite: is it the writer or some undisclosed person or the world at large that will always remember this visit?

This rule does not, of course, mean that the writer should entirely discard the passive voice, which is frequently convenient and sometimes necessary.

> The dramatists of the Restoration are little esteemed today.

> Modern readers have little esteem for the dramatists of the Restoration.

The first would be the preferred form in a paragraph on the dramatists of the Restoration, the second in a paragraph on the tastes of modern readers. The need to make a particular word the subject of the sentence will often, as in these examples, determine which voice is to be used.

The habitual use of the active voice, however, makes for forcible writing. This is true not only in narrative concerned principally with action but in writing of any kind. Many a tame sentence of description or exposition can be made lively and emphatic by substituting a transitive in the active voice for some such perfunctory expression as *there is* or *could be heard.*

There were a great number of dead leaves lying on the ground.	Dead leaves covered the ground.
At dawn the crowing of a rooster could be heard.	The cock's crow came with dawn.
The reason he left college was that his health became impaired.	Failing health compelled him to leave college.
It was not long before she was very sorry that she had said what she had.	She soon repented her words.

Note, in the examples above, that when a sentence is made stronger, it usually becomes shorter. Thus, brevity is a by-product of vigor.

15. *Put statements in positive form.*

Make definite assertions. Avoid tame, colorless, hesitating, noncommittal language. Use the word *not* as a means of denial or in antithesis, never as a means of evasion.

He was not very often on time.	He usually came late.
She did not think that studying Latin was a sensible way to use one's time.	She thought the study of Latin a waste of time.
The Taming of the Shrew is rather weak in spots. Shakespeare does not portray Katharine as a very admirable character, nor does Bianca remain long in memory as an important character in Shakespeare's works.	The women in *The Taming of the Shrew* are unattractive. Katharine is disagreeable, Bianca insignificant.

The last example, before correction, is indefinite as well as negative. The corrected version, consequently, is simply a guess at the writer's intention.

All three examples show the weakness inherent in the word *not*. Consciously or unconsciously, the reader is dissatisfied with being told only what is not; the reader wishes to be told what is. Hence, as a rule, it is better to express even a negative in positive form.

not honest	dishonest
not important	trifling
did not remember	forgot
did not pay any attention to	ignored
did not have much confidence in	distrusted

Placing negative and positive in opposition makes for a stronger structure.

Not charity, but simple justice.

Not that I loved Caesar less, but that I loved Rome more.

Ask not what your country can do for you—ask what you can do for your country.

Negative words other than *not* are usually strong.

Her loveliness I never knew / Until she smiled on me.

Statements qualified with unnecessary auxiliaries or conditionals sound irresolute.

If you would let us know the time of your arrival, we would be happy to arrange your transportation from the airport.	If you will let us know the time of your arrival, we shall be happy to arrange your transportation from the airport.
Applicants can make a good impression by being neat and punctual.	Applicants will make a good impression if they are neat and punctual.
Plath may be ranked among those modern poets who died young.	Plath was one of those modern poets who died young.

If your every sentence admits a doubt, your writing will lack authority. Save the auxiliaries *would, should, could, may, might,* and *can* for situations involving real uncertainty.

16. *Use definite, specific, concrete language.*

Prefer the specific to the general, the definite to the vague, the concrete to the abstract.

A period of unfavorable weather set in.	It rained every day for a week.
He showed satisfaction as he took possession of his well-earned reward.	He grinned as he pocketed the coin.

If those who have studied the art of writing are in accord on any one point, it is this: the surest way to arouse and hold the reader's attention is by being specific, definite, and concrete. The greatest writers—Homer, Dante, Shakespeare—are effective largely because they deal in particulars and report the details that matter. Their words call up pictures.

Jean Stafford, to cite a more modern author, demonstrates in her short story "In the Zoo" how prose is made vivid by the use of words that evoke images and sensations:

> . . . Daisy and I in time found asylum in a small menagerie down by the railroad tracks. It belonged to a gentle alcoholic ne'er-do-well, who did nothing all day long but drink bathtub gin in rickeys and play solitaire and smile to himself and talk to his animals. He had a little, stunted red vixen and a deodorized skunk, a parrot from Tahiti that spoke Parisian French, a woebegone coyote, and two capuchin monkeys, so serious and humanized, so small and sad and sweet, and so religious-looking with their tonsured heads that it was impossible not to think their gibberish was really an ordered language with a grammar that someday some philologist would understand.
>
> Gran knew about our visits to Mr. Murphy and she did not object, for it gave her keen pleasure to excoriate him when we came home. His vice was not a matter of guesswork; it was an established fact that he was half-seas over from dawn till midnight. "With the black Irish," said Gran, "the taste for drink is taken in with the mother's

milk and is never mastered. Oh, I know all about those promises to join the temperance movement and not to touch another drop. The way to Hell is paved with good intentions."*

If the experiences of Walter Mitty, of Molly Bloom, of Rabbit Angstrom have seemed for the moment real to countless readers, if in reading Faulkner we have almost the sense of inhabiting Yoknapatawpha County during the decline of the South, it is because the details used are definite, the terms concrete. It is not that every detail is given—that would be impossible, as well as to no purpose— but that all the significant details are given, and with such accuracy and vigor that readers, in imagination, can project themselves into the scene.

In exposition and in argument, the writer must likewise never lose hold of the concrete; and even when dealing with general principles, the writer must furnish particular instances of their application.

In his *Philosophy of Style,* Herbert Spencer gives two sentences to illustrate how the vague and general can be turned into the vivid and particular:

In proportion as the manners, customs, and amusements of a nation are cruel and barbarous, the regulations of its penal code will be severe.	In proportion as men delight in battles, bull-fights, and combats of gladiators, will they punish by hanging, burning, and the rack.

To show what happens when strong writing is deprived of its vigor, George Orwell once took a passage from the

*Excerpt from "In the Zoo" from *Bad Characters* by Jean Stafford. Copyright © 1964 by Jean Stafford. Copyright renewed © 1992 by Nora Cosgrove. Reprinted by permission of Farrar, Straus & Giroux, LLC. Also copyright © 1969 by Jean Stafford; reprinted by permission of Curtis Brown, Ltd.

Bible and drained it of its blood. On the left, below, is Orwell's translation; on the right, the verse from Ecclesiastes (King James Version).

Objective consideration of contemporary phenomena compels the conclusion that success or failure in competitive activities exhibits no tendency to be commensurate with innate capacity, but that a considerable element of the unpredictable must inevitably be taken into account.	I returned, and saw under the sun, that the race is not to the swift, nor the battle to the strong, neither yet bread to the wise, nor yet riches to men of understanding, nor yet favor to men of skill; but time and chance happeneth to them all.

17. *Omit needless words.*

Vigorous writing is concise. A sentence should contain no unnecessary words, a paragraph no unnecessary sentences, for the same reason that a drawing should have no unnecessary lines and a machine no unnecessary parts. This requires not that the writer make all sentences short, or avoid all detail and treat subjects only in outline, but that every word tell.

Many expressions in common use violate this principle.

the question as to whether	whether (the question whether)
there is no doubt but that	no doubt (doubtless)
used for fuel purposes	used for fuel
he is a man who	he
in a hasty manner	hastily
this is a subject that	this subject
Her story is a strange one.	Her story is strange.
the reason why is that	because

The fact that is an especially debilitating expression. It should be revised out of every sentence in which it occurs.

owing to the fact that	since (because)
in spite of the fact that	though (although)
call your attention to the fact that	remind you (notify you)
I was unaware of the fact that	I was unaware that (did not know)
the fact that he had not succeeded	his failure
the fact that I had arrived	my arrival

See also the words *case, character, nature* in Chapter IV. *Who is, which was,* and the like are often superfluous.

His cousin, who is a member of the same firm	His cousin, a member of the same firm
Trafalgar, which was Nelson's last battle	Trafalgar, Nelson's last battle

As the active voice is more concise than the passive, and a positive statement more concise than a negative one, many of the examples given under Rules 14 and 15 illustrate this rule as well.

A common way to fall into wordiness is to present a single complex idea, step by step, in a series of sentences that might to advantage be combined into one.

Macbeth was very ambitious. This led him to wish to become king of Scotland. The witches told him that this wish of his would come true. The king of Scotland at this time was Duncan. Encouraged by his wife, Macbeth murdered Duncan. He was thus enabled to succeed Duncan as king. (51 words)	Encouraged by his wife, Macbeth achieved his ambition and realized the prediction of the witches by murdering Duncan and becoming king of Scotland in his place. (26 words)

18. *Avoid a succession of loose sentences.*

This rule refers especially to loose sentences of a particular type: those consisting of two clauses, the second introduced by a conjunction or relative. A writer may err by making sentences too compact and periodic. An occasional loose sentence prevents the style from becoming too formal and gives the reader a certain relief. Consequently, loose sentences are common in easy, unstudied writing. The danger is that there may be too many of them.

An unskilled writer will sometimes construct a whole paragraph of sentences of this kind, using as connectives *and, but,* and, less frequently, *who, which, when, where,* and *while,* these last in nonrestrictive senses. (See Rule 3.)

> The third concert of the subscription series was given last evening, and a large audience was in attendance. Mr. Edward Appleton was the soloist, and the Boston Symphony Orchestra furnished the instrumental music. The former showed himself to be an artist of the first rank, while the latter proved itself fully deserving of its high reputation. The interest aroused by the series has been very gratifying to the Committee, and it is planned to give a similar series annually hereafter. The fourth concert will be given on Tuesday, May 10, when an equally attractive program will be presented.

Apart from its triteness and emptiness, the paragraph above is bad because of the structure of its sentences, with their mechanical symmetry and singsong. Compare these sentences from the chapter "What I Believe" in E. M. Forster's *Two Cheers for Democracy:*

> I believe in aristocracy, though—if that is the right word, and if a democrat may use it. Not an aristocracy of power, based upon rank and influence, but an aristocracy of the sensitive, the considerate and the plucky. Its members are to be found in all nations and classes, and all through the ages, and there is a secret understanding between them when they meet. They represent the true human tradition, the one permanent victory of our queer race over cruelty and chaos. Thousands of them perish in obscurity, a few are great names. They are sensitive for

others as well as for themselves, they are considerate without being fussy, their pluck is not swankiness but the power to endure, and they can take a joke.*

A writer who has written a series of loose sentences should recast enough of them to remove the monotony, replacing them with simple sentences, sentences of two clauses joined by a semicolon, periodic sentences of two clauses, or sentences (loose or periodic) of three clauses—whichever best represent the real relations of the thought.

19. *Express coordinate ideas in similar form.*

This principle, that of parallel construction, requires that expressions similar in content and function be outwardly similar. The likeness of form enables the reader to recognize more readily the likeness of content and function. The familiar Beatitudes exemplify the virtue of parallel construction.

> Blessed are the poor in spirit: for theirs is the kingdom of heaven.
> Blessed are they that mourn: for they shall be comforted.
> Blessed are the meek: for they shall inherit the earth.
> Blessed are they which do hunger and thirst after righteousness: for they shall be filled.

The unskilled writer often violates this principle, mistakenly believing in the value of constantly varying the form of expression. When repeating a statement to emphasize it, the writer may need to vary its form. Otherwise, the writer should follow the principle of parallel construction.

Formerly, science was taught by the textbook method, while now the laboratory method is employed.	Formerly, science was taught by the textbook method; now it is taught by the laboratory method.

The lefthand version gives the impression that the writer is undecided or timid, apparently unable or afraid to choose

*Excerpt from "What I Believe" in *Two Cheers for Democracy,* copyright 1939 and renewed 1967 by E. M. Forster, reprinted by permission of Harcourt, Inc. Also, by permission of The Provost and Scholars of King's College, Cambridge, and The Society of Authors as the literary representatives of the E. M. Forster Estate.

one form of expression and hold to it. The righthand version shows that the writer has at least made a choice and abided by it.

By this principle, an article or a preposition applying to all the members of a series must either be used only before the first term or else be repeated before each term.

the French, the Italians, Spanish, and Portuguese	the French, the Italians, the Spanish, and the Portuguese
in spring, summer, or in winter	in spring, summer, or winter (in spring, in summer, or in winter)

Some words require a particular preposition in certain idiomatic uses. When such words are joined in a compound construction, all the appropriate prepositions must be included, unless they are the same.

His speech was marked by disagreement and scorn for his opponent's position.	His speech was marked by disagreement with and scorn for his opponent's position.

Correlative expressions (*both, and; not, but; not only, but also; either, or; first, second, third;* and the like) should be followed by the same grammatical construction. Many violations of this rule can be corrected by rearranging the sentence.

It was both a long ceremony and very tedious.	The ceremony was both long and tedious.
A time not for words but action.	A time not for words but for action.
Either you must grant his request or incur his ill will.	You must either grant his request or incur his ill will.
My objections are, first, the injustice of the measure; second, that it is unconstitutional.	My objections are, first, that the measure is unjust; second, that it is unconstitutional.

It may be asked, what if you need to express a rather large number of similar ideas—say, twenty? Must you write twenty consecutive sentences of the same pattern? On closer examination, you will probably find that the difficulty is imaginary—that these twenty ideas can be classified in groups, and that you need apply the principle only within each group. Otherwise, it is best to avoid the difficulty by putting statements in the form of a table.

20. *Keep related words together.*

The position of the words in a sentence is the principal means of showing their relationship. Confusion and ambiguity result when words are badly placed. The writer must, therefore, bring together the words and groups of words that are related in thought and keep apart those that are not so related.

He noticed a large stain in the rug that was right in the center.	He noticed a large stain right in the center of the rug.
You can call your mother in London and tell her all about George's taking you out to dinner for just two dollars.	For just two dollars you can call your mother in London and tell her all about George's taking you out to dinner.
New York's first commercial human-sperm bank opened Friday with semen samples from eighteen men frozen in a stainless steel tank.	New York's first commercial human-sperm bank opened Friday when semen samples were taken from eighteen men. The samples were then frozen and stored in a stainless steel tank.

In the lefthand version of the first example, the reader has no way of knowing whether the stain was in the center of the rug or the rug was in the center of the room. In the lefthand version of the second example, the reader may well wonder which cost two dollars—the phone call or the dinner. In the lefthand version of the third example, the reader's heart goes out to those eighteen poor fellows frozen in a steel tank.

The subject of a sentence and the principal verb should not, as a rule, be separated by a phrase or clause that can be transferred to the beginning.

Toni Morrison, in *Beloved,* writes about characters who have escaped from slavery but are haunted by its heritage.	In *Beloved,* Toni Morrison writes about characters who have escaped from slavery but are haunted by its heritage.
A dog, if you fail to discipline him, becomes a household pest.	Unless disciplined, a dog becomes a household pest.

Interposing a phrase or a clause, as in the lefthand examples above, interrupts the flow of the main clause. This interruption, however, is not usually bothersome when the flow is checked only by a relative clause or by an expression in apposition. Sometimes, in periodic sentences, the interruption is a deliberate device for creating suspense. (See examples under Rule 22.)

The relative pronoun should come, in most instances, immediately after its antecedent.

There was a stir in the audience that suggested disapproval.	A stir that suggested disapproval swept the audience.
He wrote three articles about his adventures in Spain, which were published in *Harper's Magazine.*	He published three articles in *Harper's Magazine* about his adventures in Spain.
This is a portrait of Benjamin Harrison, grandson of William Henry Harrison, who became President in 1889.	This is a portrait of Benjamin Harrison, who became President in 1889. He was the grandson of William Henry Harrison.

If the antecedent consists of a group of words, the relative comes at the end of the group, unless this would cause ambiguity.

> The Superintendent of the Chicago Division, who

No ambiguity results from the above. But

> A proposal to amend the Sherman Act, which has been variously judged

leaves the reader wondering whether it is the proposal or the Act that has been variously judged. The relative clause must be moved forward, to read, "A proposal, which has been variously judged, to amend the Sherman Act...." Similarly

The grandson of William Henry Harrison, who	William Henry Harrison's grandson, Benjamin Harrison, who

A noun in apposition may come between antecedent and relative, because in such a combination no real ambiguity can arise.

> The Duke of York, his brother, who was regarded with hostility by the Whigs

Modifiers should come, if possible, next to the words they modify. If several expressions modify the same word, they should be arranged so that no wrong relation is suggested.

All the members were not present.	Not all the members were present.
She only found two mistakes.	She found only two mistakes.
The director said he hoped all members would give generously to the Fund at a meeting of the committee yesterday.	At a meeting of the committee yesterday, the director said he hoped all members would give generously to the Fund.

| Major R. E. Joyce will give a lecture on Tuesday evening in Bailey Hall, to which the public is invited on "My Experiences in Mesopotamia" at 8:00 P.M. | On Tuesday evening at eight, Major R. E. Joyce will give a lecture in Bailey Hall on "My Experiences in Mesopotamia." The public is invited. |

Note, in the last lefthand example, how swiftly meaning departs when words are wrongly juxtaposed.

21. *In summaries, keep to one tense.*

In summarizing the action of a drama, use the present tense. In summarizing a poem, story, or novel, also use the present, though you may use the past if it seems more natural to do so. If the summary is in the present tense, antecedent action should be expressed by the perfect; if in the past, by the past perfect.

> Chance prevents Friar John from delivering Friar Lawrence's letter to Romeo. Meanwhile, owing to her father's arbitrary change of the day set for her wedding, Juliet has been compelled to drink the potion on Tuesday night, with the result that Balthasar informs Romeo of her supposed death before Friar Lawrence learns of the non-delivery of the letter.

But whichever tense is used in the summary, a past tense in indirect discourse or in indirect question remains unchanged.

> The Friar confesses that it was he who married them.

Apart from the exceptions noted, the writer should use the same tense throughout. Shifting from one tense to another gives the appearance of uncertainty and irresolution.

In presenting the statements or the thought of someone else, as in summarizing an essay or reporting a speech, do not overwork such expressions as "he said," "she stated," "the speaker added," "the speaker then went on to say," "the author also thinks." Indicate clearly at the outset, once for all, that what follows is summary, and then waste no words in repeating the notification.

In notebooks, in newspapers, in handbooks of literature, summaries of one kind or another may be indispensable, and for children in primary schools retelling a story in their own words is a useful exercise. But in the criticism or interpretation of literature, be careful to avoid dropping into summary. It may be necessary to devote one or two sentences to indicating the subject, or the opening situation, of the work being discussed, or to cite numerous details to illustrate its qualities. But you should aim at writing an orderly discussion supported by evidence, not a summary with occasional comment. Similarly, if the scope of the discussion includes a number of works, as a rule it is better not to take them up singly in chronological order but to aim from the beginning at establishing general conclusions.

22. *Place the emphatic words of a sentence at the end.*

The proper place in the sentence for the word or group of words that the writer desires to make most prominent is usually the end.

Humanity has hardly advanced in fortitude since that time, though it has advanced in many other ways.	Since that time, humanity has advanced in many ways, but it has hardly advanced in fortitude.
This steel is principally used for making razors, because of its hardness.	Because of its hardness, this steel is used principally for making razors.

The word or group of words entitled to this position of prominence is usually the logical predicate—that is, the *new* element in the sentence, as it is in the second example.

The effectiveness of the periodic sentence arises from the prominence it gives to the main statement.

Four centuries ago, Christopher Columbus, one of the Italian mariners whom the decline of their own republics had put at the service of the world and of adventure, seeking for Spain a westward passage to the Indies to offset the achievement of Portuguese discoverers, lighted on America.

With these hopes and in this belief I would urge you, laying aside all hindrance, thrusting away all private aims, to devote yourself unswervingly and unflinchingly to the vigorous and successful prosecution of this war.

The other prominent position in the sentence is the beginning. Any element in the sentence other than the subject becomes emphatic when placed first.

Deceit or treachery she could never forgive.

Vast and rude, fretted by the action of nearly three thousand years, the fragments of this architecture may often seem, at first sight, like works of nature.

Home is the sailor.

A subject coming first in its sentence may be emphatic, but hardly by its position alone. In the sentence

Great kings worshiped at his shrine

the emphasis upon *kings* arises largely from its meaning and from the context. To receive special emphasis, the subject of a sentence must take the position of the predicate.

Through the middle of the valley flowed a winding stream.

The principle that the proper place for what is to be made most prominent is the end applies equally to the words of a sentence, to the sentences of a paragraph, and to the paragraphs of a composition.

III

A Few Matters of Form

Colloquialisms.   If you use a colloquialism or a slang word or phrase, simply use it; do not draw attention to it by enclosing it in quotation marks. To do so is to put on airs, as though you were inviting the reader to join you in a select society of those who know better.

Exclamations.   Do not attempt to emphasize simple statements by using a mark of exclamation.

It was a wonderful show!   It was a wonderful show.

The exclamation mark is to be reserved for use after true exclamations or commands.

What a wonderful show!
Halt!

Headings.   If a manuscript is to be submitted for publication, leave plenty of space at the top of page 1. The editor will need this space to write directions to the compositor. Place the heading, or title, at least a fourth of the way down the page. Leave a blank line, or its equivalent in space, after the heading. On succeeding pages, begin near the top, but not so near as to give a crowded appearance. Omit the period after a title or heading. A question mark or an exclamation point may be used if the heading calls for it.

Hyphen.   When two or more words are combined to form a compound adjective, a hyphen is usually required.

"He belonged to the leisure class and enjoyed leisure-class pursuits." "She entered her boat in the round-the-island race."

Do not use a hyphen between words that can better be written as one word: *water-fowl, waterfowl.* Common sense will aid you in the decision, but a dictionary is more reliable. The steady evolution of the language seems to favor union: two words eventually become one, usually after a period of hyphenation.

bed chamber	bed-chamber	bedchamber
wild life	wild-life	wildlife
bell boy	bell-boy	bellboy

The hyphen can play tricks on the unwary, as it did in Chattanooga when two newspapers merged—the *News* and the *Free Press.* Someone introduced a hyphen into the merger, and the paper became *The Chattanooga News-Free Press,* which sounds as though the paper were news-free, or devoid of news. Obviously, we ask too much of a hyphen when we ask it to cast its spell over words it does not adjoin.

Margins.   Keep righthand and lefthand margins roughly the same width. Exception: If a great deal of annotating or editing is anticipated, the lefthand margin should be roomy enough to accommodate this work.

Numerals.   Do not spell out dates or other serial numbers. Write them in figures or in Roman notation, as appropriate.

August 9, 1988	Part XII
Rule 3	352d Infantry

Exception: When they occur in dialogue, most dates and numbers are best spelled out.

"I arrived home on August ninth."

"In the year 1990, I turned twenty-one."

"Read Chapter Twelve."

Parentheses.   A sentence containing an expression in parentheses is punctuated outside the last mark of parenthesis exactly as if the parenthetical expression were absent. The expression within the marks is punctuated as if it stood by itself, except that the final stop is omitted unless it is a question mark or an exclamation point.

> I went to her house yesterday (my third attempt to see her), but she had left town.

> He declares (and why should we doubt his good faith?) that he is now certain of success.

(When a wholly detached expression or sentence is parenthesized, the final stop comes before the last mark of parenthesis.)

Quotations.   Formal quotations cited as documentary evidence are introduced by a colon and enclosed in quotation marks.

> The United States Coast Pilot has this to say of the place: "Bracy Cove, 0.5 mile eastward of Bear Island, is exposed to southeast winds, has a rocky and uneven bottom, and is unfit for anchorage."

A quotation grammatically in apposition or the direct object of a verb is preceded by a comma and enclosed in quotation marks.

> I am reminded of the advice of my neighbor, "Never worry about your heart till it stops beating."

> Mark Twain says, "A classic is something that everybody wants to have read and nobody wants to read."

When a quotation is followed by an attributive phrase, the comma is enclosed within the quotation marks.

> "I can't attend," she said.

Typographical usage dictates that the comma be inside the marks, though logically it often seems not to belong there.

> "The Fish," "Poetry," and "The Monkeys" are in Marianne Moore's *Selected Poems.*

When quotations of an entire line, or more, of either verse or prose are to be distinguished typographically from text matter, as are the quotations in this book, begin on a fresh line and indent. Quotation marks should not be used unless they appear in the original, as in dialogue.

> Wordsworth's enthusiasm for the French Revolution was at first unbounded:
>
>> Bliss was it in that dawn to be alive,
>> But to be young was very heaven!

Quotations introduced by *that* are indirect discourse and not enclosed in quotation marks.

> Keats declares that beauty is truth, truth beauty.
>
> Dickinson states that a coffin is a small domain.

Proverbial expressions and familiar phrases of literary origin require no quotation marks.

> These are the times that try men's souls.
>
> He lives far from the madding crowd.

References.   In scholarly work requiring exact references, abbreviate titles that occur frequently, giving the full forms in an alphabetical list at the end. As a general practice, give the references in parentheses or in footnotes, not in the body of the sentence. Omit the words *act, scene, line, book, volume, page,* except when referring to only one of them. Punctuate as indicated below.

in the second scene of the third act	in III.ii (Better still, simply insert III.ii in parentheses at the proper place in the sentence.)

After the killing of Polonius, Hamlet is placed under guard (IV.ii.14).

2 Samuel i:17–27

Othello II.iii. 264–267, III.iii.155–161

Syllabication.   When a word must be divided at the end of a line, consult a dictionary to learn the syllables between which division should be made. The student will do well to examine the syllable division in a number of pages of any carefully printed book.

Titles.   For the titles of literary works, scholarly usage prefers italics with capitalized initials. The usage of editors and publishers varies, some using italics with capitalized initials, others using Roman with capitalized initials and with or without quotation marks. Use italics (indicated in manuscript by underscoring) except in writing for a periodical that follows a different practice. Omit initial *A* or *The* from titles when you place the possessive before them.

A Tale of Two Cities; Dickens's *Tale of Two Cities.*

The Age of Innocence; Wharton's *Age of Innocence.*

IV

Words and Expressions Commonly Misused

MANY of the words and expressions listed here are not so much bad English as bad style, the commonplaces of careless writing. As illustrated under *Feature*, the proper correction is likely to be not the replacement of one word or set of words by another but the replacement of vague generality by definite statement.

The shape of our language is not rigid; in questions of usage we have no lawgiver whose word is final. Students whose curiosity is aroused by the interpretations that follow, or whose doubts are raised, will wish to pursue their investigations further. Books useful in such pursuits are *Merriam Webster's Collegiate Dictionary*, Tenth Edition; *The American Heritage Dictionary of the English Language*, Third Edition; *Webster's Third New International Dictionary; The New Fowler's Modern English Usage*, Third Edition, edited by R. W. Burchfield; *Modern American Usage: A Guide* by Wilson Follett and Erik Wensberg; and *The Careful Writer* by Theodore M. Bernstein.

Aggravate. Irritate.   The first means "to add to" an already troublesome or vexing matter or condition. The second means "to vex" or "to annoy" or "to chafe."

All right.   Idiomatic in familiar speech as a detached phrase in the sense "Agreed," or "Go ahead," or "O.K." Properly written as two words—*all right*.

Allude. Do not confuse with *elude*. You *allude* to a book; you *elude* a pursuer. Note, too, that *allude* is not synonymous with *refer*. An allusion is an indirect mention, a reference is a specific one.

Allusion. Easily confused with *illusion*. The first means "an indirect reference"; the second means "an unreal image" or "a false impression."

Alternate. Alternative. The words are not always interchangeable as nouns or adjectives. The first means every other one in a series; the second, one of two possibilities. As the other one of a series of two, an *alternate* may stand for "a substitute," but an *alternative,* although used in a similar sense, connotes a matter of choice that is never present with *alternate.*

> As the flooded road left them no alternative, they took the alternate route.

Among. Between. When more than two things or persons are involved, *among* is usually called for: "The money was divided among the four players." When, however, more than two are involved but each is considered individually, *between* is preferred: "an agreement between the six heirs."

And/or. A device, or shortcut, that damages a sentence and often leads to confusion or ambiguity.

First of all, would an honor system successfully cut down on the amount of stealing and/or cheating?	First of all, would an honor system reduce the incidence of stealing or cheating or both?

Anticipate. Use *expect* in the sense of simple expectation.

I anticipated that he would look older.	I expected that he would look older.
My brother anticipated the upturn in the market.	My brother expected the upturn in the market.

In the second example, the word *anticipated* is ambiguous. It could mean simply that the brother believed the upturn would occur, or it could mean that he acted in advance of the expected upturn—by buying stock, perhaps.

Anybody. In the sense of "any person," not to be written as two words. *Any body* means "any corpse," or "any human form," or "any group." The rule holds equally for *everybody, nobody,* and *somebody.*

Anyone. In the sense of "anybody," written as one word. *Any one* means "any single person" or "any single thing."

As good or better than. Expressions of this type should be corrected by rearranging the sentences.

My opinion is as good or better than his.	My opinion is as good as his, or better (if not better).

As to whether. *Whether* is sufficient.

As yet. *Yet* nearly always is as good, if not better.

No agreement has been reached as yet.	No agreement has yet been reached.

The chief exception is at the beginning of a sentence, where *yet* means something different.

Yet (*or* despite everything) he has not succeeded.

As yet (*or* so far) he has not succeeded.

Being. Not appropriate after *regard . . . as.*

He is regarded as being the best dancer in the club.	He is regarded as the best dancer in the club.

But. Unnecessary after *doubt* and *help.*

I have no doubt but that	I have no doubt that
He could not help but see that	He could not help seeing that

The too-frequent use of *but* as a conjunction leads to the fault discussed under Rule 18. A loose sentence formed with *but* can usually be converted into a periodic sentence formed with *although*.

Particularly awkward is one *but* closely following another, thus making a contrast to a contrast, or a reservation to a reservation. This is easily corrected by rearrangement.

Our country had vast resources but seemed almost wholly unprepared for war. But within a year it had created an army of four million.	Our country seemed almost wholly unprepared for war, but it had vast resources. Within a year it had created an army of four million.

Can.   Means "am (is, are) able." Not to be used as a substitute for *may*.

Care less.   The dismissive "I couldn't care less" is often used with the shortened "not" mistakenly (and mysteriously) omitted: "I could care less." The error destroys the meaning of the sentence and is careless indeed.

Case.   Often unnecessary.

In many cases, the rooms lacked air conditioning.	Many of the rooms lacked air conditioning.
It has rarely been the case that any mistake has been made.	Few mistakes have been made.

Certainly.   Used indiscriminately by some speakers, much as others use *very*, in an attempt to intensify any and every statement. A mannerism of this kind, bad in speech, is even worse in writing.

Character.   Often simply redundant, used from a mere habit of wordiness.

acts of a hostile character	hostile acts

Claim (verb).   With object-noun, means "lay claim to." May be used with a dependent clause if this sense is clearly intended: "She claimed that she was the sole heir." (But even here *claimed to be* would be better.) Not to be used as a substitute for *declare, maintain,* or *charge.*

| He claimed he knew how. | He declared he knew how. |

Clever.   Note that the word means one thing when applied to people, another when applied to horses. A clever horse is a good-natured one, not an ingenious one.

Compare.   To *compare to* is to point out or imply resemblances between objects regarded as essentially of a different order; to *compare with* is mainly to point out differences between objects regarded as essentially of the same order. Thus, life has been *compared to* a pilgrimage, *to* a drama, *to* a battle; Congress may be *compared with* the British Parliament. Paris has been *compared to* ancient Athens; it may be *compared with* modern London.

Comprise.   Literally, "embrace": A zoo comprises mammals, reptiles, and birds (because it "embraces," or "includes," them). But animals do not comprise ("embrace") a zoo—they constitute a zoo.

Consider.   Not followed by *as* when it means "believe to be."

| I consider him as competent. | I consider him competent. |

When *considered* means "examined" or "discussed," it is followed by *as:*

The lecturer considered Eisenhower first as soldier and second as administrator.

Contact.   As a transitive verb, the word is vague and self-important. Do not *contact* people; get in touch with them, look them up, phone them, find them, or meet them.

Cope.   An intransitive verb used with *with*. In formal writing, one doesn't "cope," one "copes with" something or somebody.

I knew they'd cope. (jocular)	I knew they would cope with the situation.

Currently.   In the sense of *now* with a verb in the present tense, *currently* is usually redundant; emphasis is better achieved through a more precise reference to time.

We are currently reviewing your application.	We are at this moment reviewing your application.

Data.   Like *strata, phenomena,* and *media, data* is a plural and is best used with a plural verb. The word, however, is slowly gaining acceptance as a singular.

The data is misleading.	These data are misleading.

Different than.   Here logic supports established usage: one thing differs *from* another, hence, *different from.* Or, *other than, unlike.*

Disinterested.   Means "impartial." Do not confuse it with *uninterested,* which means "not interested in."

Let a disinterested person judge our dispute. (an impartial person)

This man is obviously uninterested in our dispute. (couldn't care less)

Divided into.   Not to be misused for *composed of.* The line is sometimes difficult to draw; doubtless plays are divided into acts, but poems are composed of stanzas. An apple, halved, is divided into sections, but an apple is composed of seeds, flesh, and skin.

Due to.   Loosely used for *through, because of,* or *owing to,* in adverbial phrases.

He lost the first game due to carelessness.	He lost the first game because of carelessness.

In correct use, synonymous with *attributable to:* "The accident was due to bad weather"; "losses due to preventable fires."

Each and every one.   Pitchman's jargon. Avoid, except in dialogue.

It should be a lesson to each and every one of us.	It should be a lesson to every one of us (to us all).

Effect.   As a noun, means "result"; as a verb, means "to bring about," "to accomplish" (not to be confused with *affect,* which means "to influence").

As a noun, often loosely used in perfunctory writing about fashions, music, painting, and other arts: "a Southwestern effect"; "effects in pale green"; "very delicate effects"; "subtle effects"; "a charming effect was produced." The writer who has a definite meaning to express will not take refuge in such vagueness.

Enormity.   Use only in the sense of "monstrous wickedness." Misleading, if not wrong, when used to express bigness.

Enthuse.   An annoying verb growing out of the noun *enthusiasm.* Not recommended.

She was enthused about her new car.	She was enthusiastic about her new car.
She enthused about her new car.	She talked enthusiastically (expressed enthusiasm) about her new car.

Etc.   Literally, "and other things"; sometimes loosely used to mean "and other persons." The phrase is equivalent to *and the rest, and so forth,* and hence is not to be used if one of these would be insufficient—that is, if the reader would be left in doubt as to any important particulars. Least open to objection when it represents the last terms of a list already given almost in full, or immaterial words at the end of a quotation.

At the end of a list introduced by *such as, for example,* or any similar expression, *etc.* is incorrect. In formal writing, *etc.* is a misfit. An item important enough to call for *etc.* is probably important enough to be named.

Fact. Use this word only of matters capable of direct verification, not of matters of judgment. That a particular event happened on a given date and that lead melts at a certain temperature are facts. But such conclusions as that Napoleon was the greatest of modern generals or that the climate of California is delightful, however defensible they may be, are not properly called facts.

Facility. Why must jails, hospitals, and schools suddenly become "facilities"?

Parents complained bitterly about the fire hazard in the wooden facility.	Parents complained bitterly about the fire hazard in the wooden schoolhouse.
He has been appointed warden of the new facility.	He has been appointed warden of the new prison.

Factor. A hackneyed word; the expressions of which it is a part can usually be replaced by something more direct and idiomatic.

Her superior training was the great factor in her winning the match.	She won the match by being better trained.
Air power is becoming an increasingly important factor in deciding battles.	Air power is playing a larger and larger part in deciding battles.

Farther. Further. The two words are commonly interchanged, but there is a distinction worth observing: *farther* serves best as a distance word, *further* as a time or quantity word. You chase a ball *farther* than the other fellow; you pursue a subject *further.*

Feature.   Another hackneyed word; like *factor,* it usually adds nothing to the sentence in which it occurs.

A feature of the entertainment especially worthy of mention was the singing of Allison Jones.	(Better use the same number of words to tell what Allison Jones sang and how she sang it.)

As a verb, in the sense of "offer as a special attraction," it is to be avoided.

Finalize.   A pompous, ambiguous verb. (See Chapter V, Reminder 21.)

Fix.   Colloquial in America for *arrange, prepare, mend.* The usage is well established. But bear in mind that this verb is from *figere:* "to make firm," "to place definitely." These are the preferred meanings of the word.

Flammable.   An oddity, chiefly useful in saving lives. The common word meaning "combustible" is *inflammable.* But some people are thrown off by the *in-* and think *inflammable* means "not combustible." For this reason, trucks carrying gasoline or explosives are now marked FLAMMABLE. Unless you are operating such a truck and hence are concerned with the safety of children and illiterates, use *inflammable.*

Folk.   A collective noun, equivalent to *people.* Use the singular form only. *Folks,* in the sense of "parents," "family," "those present," is colloquial and too folksy for formal writing.

Her folks arrived by the afternoon train.	Her father and mother arrived by the afternoon train.

Fortuitous.   Limited to what happens by chance. Not to be used for *fortunate* or *lucky.*

Get.   The colloquial *have got* for *have* should not be used in writing. The preferable form of the participle is *got,* not *gotten.*

He has not got any sense.	He has no sense.
They returned without having gotten any.	They returned without having got any.

Gratuitous.   Means "unearned," or "unwarranted."

The insult seemed gratuitous. (undeserved)

He is a man who.   A common type of redundant expression; see Rule 17.

He is a man who is very ambitious.	He is very ambitious.
Vermont is a state that attracts visitors because of its winter sports.	Vermont attracts visitors because of its winter sports.

Hopefully.   This once-useful adverb meaning "with hope" has been distorted and is now widely used to mean "I hope" or "it is to be hoped." Such use is not merely wrong, it is silly. To say, "Hopefully I'll leave on the noon plane" is to talk nonsense. Do you mean you'll leave on the noon plane in a hopeful frame of mind? Or do you mean you hope you'll leave on the noon plane? Whichever you mean, you haven't said it clearly. Although the word in its new, free-floating capacity may be pleasurable and even useful to many, it offends the ear of many others, who do not like to see words dulled or eroded, particularly when the erosion leads to ambiguity, softness, or nonsense.

However.   Avoid starting a sentence with *however* when the meaning is "nevertheless." The word usually serves better when not in first position.

| The roads were almost impassable. However, we at last succeeded in reaching camp. | The roads were almost impassable. At last, however, we succeeded in reaching camp. |

When *however* comes first, it means "in whatever way" or "to whatever extent."

However you advise him, he will probably do as he thinks best.

However discouraging the prospect, they never lost heart.

Illusion.   See *allusion.*

Imply. Infer.   Not interchangeable. Something implied is something suggested or indicated, though not expressed. Something inferred is something deduced from evidence at hand.

Farming implies early rising.

Since she was a farmer, we inferred that she got up early.

Importantly.   Avoid by rephrasing.

| More importantly, he paid for the damages. | What's more, he paid for the damages. |
| With the breeze freshening, he altered course to pass inside the island. More importantly, as things turned out, he tucked in a reef. | With the breeze freshening, he altered course to pass inside the island. More important, as things turned out, he tucked in a reef. |

In regard to.   Often wrongly written *in regards to.* But *as regards* is correct, and means the same thing.

In the last analysis.   A bankrupt expression.

Inside of. Inside.   The *of* following *inside* is correct in the adverbial meaning "in less than." In other meanings, *of* is unnecessary.

Inside of five minutes I'll be inside the bank.

Insightful. The word is a suspicious overstatement for "perceptive." If it is to be used at all, it should be used for instances of remarkably penetrating vision. Usually, it crops up merely to inflate the commonplace.

That was an insightful remark you made.	That was a perceptive remark you made.

In terms of. A piece of padding usually best omitted.

The job was unattractive in terms of salary.	The salary made the job unattractive.

Interesting. An unconvincing word; avoid it as a means of introduction. Instead of announcing that what you are about to tell is interesting, make it so.

An interesting story is told of	(Tell the story without preamble.)
In connection with the forthcoming visit of Mr. B. to America, it is interesting to recall that he	Mr. B., who will soon visit America

Also to be avoided in introduction is the word *funny*. Nothing becomes funny by being labeled so.

Irregardless. Should be *regardless*. The error results from failure to see the negative in *-less* and from a desire to get it in as a prefix, suggested by such words as *irregular, irresponsible,* and, perhaps especially, *irrespective*.

-ize. Do not coin verbs by adding this tempting suffix. Many good and useful verbs do end in *-ize: summarize, fraternize, harmonize, fertilize*. But there is a growing list of abominations: *containerize, prioritize, finalize,* to name three. Be suspicious of *-ize;* let your ear and your eye guide you. Never tack *-ize* onto a noun to create a verb. Usually you will discover that a useful verb already exists. Why say "utilize" when there is the simple, unpretentious word *use*?

Kind of. Except in familiar style, not to be used as a substitute for *rather* or *something like*. Restrict it to its literal sense: "Amber is a kind of fossil resin"; "I dislike that kind of publicity." The same holds true for *sort of*.

Lay. A transitive verb. Except in slang ("Let it lay"), do not misuse it for the intransitive verb *lie*. The hen, or the play, *lays* an egg; the llama *lies* down. The playwright went home and *lay* down.

lie, lay, lain, lying

lay, laid, laid, laying

Leave. Not to be misused for *let*.

Leave it stand the way it is.	Let it stand the way it is.
Leave go of that rope!	Let go of that rope!

Less. Should not be misused for *fewer*.

They had less workers than in the previous campaign.	They had fewer workers than in the previous campaign.

Less refers to quantity, *fewer* to number. "His troubles are less than mine" means "His troubles are not so great as mine." "His troubles are fewer than mine" means "His troubles are not so numerous as mine."

Like. Not to be used for the conjunction *as*. *Like* governs nouns and pronouns; before phrases and clauses the equivalent word is *as*.

We spent the evening like in the old days.	We spent the evening as in the old days.
Chloë smells good, like a baby should.	Chloë smells good, as a baby should.

The use of *like* for *as* has its defenders; they argue that any usage that achieves currency becomes valid automatically. This, they say, is the way the language is formed. It is and it isn't. An expression sometimes merely enjoys a vogue, much as an article of apparel does. *Like* has long been widely misused by the illiterate; lately it has been taken up by the knowing and the well-informed, who find it catchy, or liberating, and who use it as though they were slumming. If every word or device that achieved currency were immediately authenticated, simply on the ground of popularity, the language would be as chaotic as a ball game with no foul lines. For the student, perhaps the most useful thing to know about *like* is that most carefully edited publications regard its use before phrases and clauses as simple error.

Line. Along these lines.   *Line* in the sense of "course of procedure, conduct, thought" is allowable but has been so overworked, particularly in the phrase *along these lines,* that a writer who aims at freshness or originality had better discard it entirely.

Mr. B. also spoke along the same lines.	Mr. B. also spoke to the same effect.
She is studying along the line of French literature.	She is studying French literature.

Literal. Literally.   Often incorrectly used in support of exaggeration or violent metaphor.

a literal flood of abuse	a flood of abuse
literally dead with fatigue	almost dead with fatigue

Loan.   A noun. As a verb, prefer *lend.*

Lend me your ears.

the loan of your ears

Meaningful.   A bankrupt adjective. Choose another, or rephrase.

His was a meaningful contribution.	His contribution counted heavily.
We are instituting many meaningful changes in the curriculum.	We are improving the curriculum in many ways.

Memento.   Often incorrectly written *momento*.

Most.   Not to be used for *almost* in formal composition.

most everybody	almost everybody
most all the time	almost all the time

Nature.   Often simply redundant, used like *character*.

acts of a hostile nature	hostile acts

Nature should be avoided in such vague expressions as "a lover of nature," "poems about nature." Unless more specific statements follow, the reader cannot tell whether the poems have to do with natural scenery, rural life, the sunset, the untracked wilderness, or the habits of squirrels.

Nauseous. Nauseated.   The first means "sickening to contemplate"; the second means "sick at the stomach." Do not, therefore, say, "I feel nauseous," unless you are sure you have that effect on others.

Nice.   A shaggy, all-purpose word, to be used sparingly in formal composition. "I had a nice time." "It was nice weather." "She was so nice to her mother." The meanings are indistinct. *Nice* is most useful in the sense of "precise" or "delicate": "a nice distinction."

Nor.   Often used wrongly for *or* after negative expressions.

He cannot eat nor sleep.

He cannot eat or sleep.

He can neither eat nor sleep.

He cannot eat nor can he sleep.

Noun used as verb. Many nouns have lately been pressed into service as verbs. Not all are bad, but all are suspect.

Be prepared for kisses when you gift your girlfriend with this merry scent.

Be prepared for kisses when you give your girlfriend this merry scent.

The candidate hosted a dinner for fifty of her workers.

The candidate gave a dinner for fifty of her workers.

The meeting was chaired by Mr. Oglethorp.

Mr. Oglethorp was chair of the meeting.

She headquarters in Newark.

She has headquarters in Newark.

The theater troupe debuted last fall.

The theatre troupe made its debut last fall.

Offputting. Ongoing. Newfound adjectives, to be avoided because they are inexact and clumsy. *Ongoing* is a mix of "continuing" and "active" and is usually superfluous.

He devoted all his spare time to the ongoing program for aid to the elderly.

He devoted all his spare time to the program for aid to the elderly.

Offputting might mean "objectionable," "disconcerting," "distasteful." Select instead a word whose meaning is clear. As a simple test, transform the participles to verbs. It is possible to *upset* something. But to *offput?* To *ongo?*

One. In the sense of "a person," not to be followed by *his* or *her*.

One must watch his step.	One must watch one's step. (You must watch your step.)

One of the most. Avoid this feeble formula. "One of the most exciting developments of modern science is . . ."; "Switzerland is one of the most beautiful countries of Europe." There is nothing wrong with the grammar; the formula is simply threadbare.

-oriented. A clumsy, pretentious device, much in vogue. Find a better way of indicating orientation or alignment or direction.

It was a manufacturing-oriented company.	It was a company chiefly concerned with manufacturing.
Many of the skits are situation-oriented.	Many of the skits rely on situation.

Partially. Not always interchangeable with *partly.* Best used in the sense of "to a certain degree," when speaking of a condition or state: "I'm partially resigned to it." *Partly* carries the idea of a part as distinct from the whole—usually a physical object.

The log was partially submerged.	The log was partly submerged.
She was partially in and partially out.	She was partly in and partly out. She was part in, part out.

Participle for verbal noun.

There was little prospect of the Senate accepting even this compromise.	There was little prospect of the Senate's accepting even this compromise.

In the lefthand column, *accepting* is a present participle; in the righthand column, it is a verbal noun (gerund). The construction shown in the lefthand column is occasionally found, and has its defenders. Yet it is easy to see that the second sentence has to do not with a prospect of the Senate but with a prospect of accepting.

Any sentence in which the use of the possessive is awkward or impossible should of course be recast.

In the event of a reconsideration of the whole matter's becoming necessary	If it should become necessary to reconsider the whole matter
There was great dissatisfaction with the decision of the arbitrators being favorable to the company.	There was great dissatisfaction with the arbitrators' decision in favor of the company.

People.   A word with many meanings. (*The American Heritage Dictionary,* Third Edition, gives nine.) *The people* is a political term, not to be confused with *the public.* From the people comes political support or opposition; from the public comes artistic appreciation or commercial patronage.

The word *people* is best not used with words of number, in place of *persons.* If of "six people" five went away, how many people would be left? Answer: one people.

Personalize.   A pretentious word, often carrying bad advice. Do not *personalize* your prose; simply make it good and keep it clean. See Chapter V, Reminder 1.

a highly personalized affair	a highly personal affair
Personalize your stationery.	Design a letterhead.

Personally.   Often unnecessary.

Personally, I thought it was a good book.	I thought it a good book.

Possess.   Often used because to the writer it sounds more impressive than *have* or *own.* Such usage is not incorrect but is to be guarded against.

She possessed great courage.	She had great courage (was very brave).
He was the fortunate possessor of	He was lucky enough to own

Presently. Has two meanings: "in a short while" and "currently." Because of this ambiguity it is best restricted to the first meaning: "She'll be here presently" ("soon," or "in a short time").

Prestigious. Often an adjective of last resort. It's in the dictionary, but that doesn't mean you have to use it.

Refer. See *allude.*

Regretful. Sometimes carelessly used for *regrettable:* "The mixup was due to a regretful breakdown in communications."

Relate. Not to be used intransitively to suggest rapport.

I relate well to Janet.	Janet and I see things the same way.
	Janet and I have a lot in common.

Respective. Respectively. These words may usually be omitted with advantage.

Works of fiction are listed under the names of their respective authors.	Works of fiction are listed under the names of their authors.
The mile run and the two-mile run were won by Jones and Cummings respectively.	The mile run was won by Jones, the two-mile run by Cummings.

Secondly, thirdly, etc. Unless you are prepared to begin with *firstly* and defend it (which will be difficult), do not prettify numbers with *-ly.* Modern usage prefers *second, third,* and so on.

Shall. Will. In formal writing, the future tense requires *shall* for the first person, *will* for the second and third. The formula to express the speaker's belief regarding a future action or state is *I shall; I will* expresses determination or consent. A swimmer in distress cries, "I shall drown; no one will save me!" A suicide puts it the other way: "I will drown; no one shall save me!" In relaxed speech, however, the words *shall* and *will* are seldom used precisely; our ear guides us or fails to guide us, as the case may be, and we are quite likely to drown when we want to survive and survive when we want to drown.

So. Avoid, in writing, the use of *so* as an intensifier: "so good"; "so warm"; "so delightful."

Sort of. See *kind of.*

Split infinitive. There is precedent from the fourteenth century down for interposing an adverb between *to* and the infinitive it governs, but the construction should be avoided unless the writer wishes to place unusual stress on the adverb.

to diligently inquire   to inquire diligently

For another side to the split infinitive, see Chapter V, Reminder 14.

State. Not to be used as a mere substitute for *say, remark.* Restrict it to the sense of "express fully or clearly": "He refused to state his objections."

Student body. Nine times out of ten a needless and awkward expression, meaning no more than the simple word *students.*

a member of the student   a student
body

popular with the student   liked by the students
body

Than.   Any sentence with *than* (to express comparison) should be examined to make sure no essential words are missing.

I'm probably closer to my mother than my father. (Ambiguous.)	I'm probably closer to my mother than to my father.
	I'm probably closer to my mother than my father is.
It looked more like a cormorant than a heron.	It looked more like a cormorant than like a heron.

Thanking you in advance.   This sounds as if the writer meant, "It will not be worth my while to write to you again." In making your request, write "Will you please," or "I shall be obliged." Then, later, if you feel moved to do so, or if the circumstances call for it, write a letter of acknowledgment.

That. Which.   *That* is the defining, or restrictive, pronoun, *which* the nondefining, or nonrestrictive. (See Rule 3.)

The lawn mower that is broken is in the garage. (Tells which one.)

The lawn mower, which is broken, is in the garage. (Adds a fact about the only mower in question.)

The use of *which* for *that* is common in written and spoken language ("Let us now go even unto Bethlehem, and see this thing which is come to pass."). Occasionally *which* seems preferable to *that*, as in the sentence from the Bible. But it would be a convenience to all if these two pronouns were used with precision. Careful writers, watchful for small conveniences, go *which*-hunting, remove the defining *whiches*, and by so doing improve their work.

The foreseeable future.   A cliché, and a fuzzy one. How much of the future is foreseeable? Ten minutes? Ten years? Any of it? By whom is it foreseeable? Seers? Experts? Everybody?

The truth is. . . . The fact is. . . . A bad beginning for a sentence. If you feel you are possessed of the truth, or of the fact, simply state it. Do not give it advance billing.

They. He or She. Do not use *they* when the antecedent is a distributive expression such as *each, each one, everybody, every one, many a man.* Use the singular pronoun.

Every one of us knows they are fallible.	Every one of us knows he is fallible.
Everyone in the community, whether they are a member of the Association or not, is invited to attend.	Everyone in the community, whether he is a member of the Association or not, is invited to attend.

A similar fault is the use of the plural pronoun with the antecedent *anybody, somebody, someone,* the intention being either to avoid the awkward *he or she* or to avoid committing oneself to one or the other. Some bashful speakers even say, "A friend of mine told me that they. . . ."

The use of *he* as a pronoun for nouns embracing both genders is a simple, practical convention rooted in the beginnings of the English language. Currently, however, many writers find the use of the generic *he* or *his* to rename indefinite antecedents limiting or offensive. Substituting *he or she* in its place is the logical thing to do if it works. But it often doesn't work, if only because repetition makes it sound boring or silly.

Consider these strategies to avoid an awkward overuse of *he or she* or an unintentional emphasis on the masculine:

Use the plural rather than the singular.

The writer must address his readers' concerns.	Writers must address their readers' concerns.

Eliminate the pronoun altogether.

The writer must address his readers' concerns.	The writer must address readers' concerns.

Substitute the second person for the third person.

| The writer must address his readers' concerns. | As a writer, you must address your readers' concerns. |

No one need fear to use *he* if common sense supports it. If you think *she* is a handy substitute for *he,* try it and see what happens. Alternatively, put all controversial nouns in the plural and avoid the choice of sex altogether, although you may find your prose sounding general and diffuse as a result.

This. The pronoun *this,* referring to the complete sense of a preceding sentence or clause, can't always carry the load and so may produce an imprecise statement.

| Visiting dignitaries watched yesterday as ground was broken for the new high-energy physics laboratory with a blowout safety wall. This is the first visible evidence of the university's plans for modernization and expansion. | Visiting dignitaries watched yesterday as ground was broken for the new high-energy physics laboratory with a blowout safety wall. The ceremony afforded the first visible evidence of the university's plans for modernization and expansion. |

In the lefthand example above, *this* does not immediately make clear what the first visible evidence is.

Thrust. This showy noun, suggestive of power, hinting of sex, is the darling of executives, politicos, and speechwriters. Use it sparingly. Save it for specific application.

| Our reorganization plan has a tremendous thrust. | The piston has a five-inch thrust. |
| The thrust of his letter was that he was working more hours than he'd bargained for. | The point he made in his letter was that he was working more hours than he'd bargained for. |

Tortuous. Torturous. A winding road is *tortuous,* a painful ordeal is *torturous.* Both words carry the idea of "twist," the twist having been a form of torture.

Transpire. Not to be used in the sense of "happen," "come to pass." Many writers so use it (usually when groping toward imagined elegance), but their usage finds little support in the Latin "breathe across or through." It is correct, however, in the sense of "become known." "Eventually, the grim account of his villainy transpired" (literally, "leaked through or out").

Try. Takes the infinitive: "try to mend it," not "try and mend it." Students of the language will argue that *try and* has won through and become idiom. Indeed it has, and it is relaxed and acceptable. But *try to* is precise, and when you are writing formal prose, try and write *try to.*

Type. Not a synonym for *kind of.* The examples below are common vulgarisms.

that type employee	that kind of employee
I dislike that type publicity.	I dislike that kind of publicity.
small, home-type hotels	small, homelike hotels
a new type plane	a plane of a new design (new kind)

Unique. Means "without like or equal." Hence, there can be no degrees of uniqueness.

It was the most unique coffee maker on the market.	It was a unique coffee maker.
The balancing act was very unique.	The balancing act was unique.
Of all the spiders, the one that lives in a bubble under water is the most unique.	Among spiders, the one that lives in a bubble under water is unique.

Utilize.   Prefer *use.*

I utilized the facilities.	I used the toilet.
He utilized the dish-washer.	He used the dishwasher.

Verbal.   Sometimes means "word for word" and in this sense may refer to something expressed in writing. *Oral* (from Latin *ōs,* "mouth") limits the meaning to what is transmitted by speech. *Oral agreement* is more precise than *verbal agreement.*

Very.   Use this word sparingly. Where emphasis is necessary, use words strong in themselves.

While.   Avoid the indiscriminate use of this word for *and, but,* and *although.* Many writers use it frequently as a substitute for *and* or *but,* either from a mere desire to vary the connective or from doubt about which of the two connectives is more appropriate. In this use it is best replaced by a semicolon.

The office and salesrooms are on the ground floor, while the rest of the building is used for manufacturing.	The office and salesrooms are on the ground floor; the rest of the building is used for manufacturing.

Its use as a virtual equivalent of *although* is allowable in sentences where this leads to no ambiguity or absurdity.

> While I admire his energy, I wish it were employed in a better cause.

This is entirely correct, as shown by the paraphrase

> I admire his energy; at the same time, I wish it were employed in a better cause.

Compare:

> While the temperature reaches 90 or 95 degrees in the daytime, the nights are often chilly.

The paraphrase shows why the use of *while* is incorrect:

> The temperature reaches 90 or 95 degrees in the day-time; at the same time the nights are often chilly.

In general, the writer will do well to use *while* only with strict literalness, in the sense of "during the time that."

-wise. Not to be used indiscriminately as a pseudosuffix: *taxwise, pricewise, marriagewise, prosewise, saltwater taffywise.* Chiefly useful when it means "in the manner of": *clockwise.* There is not a noun in the language to which *-wise* cannot be added if the spirit moves one to add it. The sober writer will abstain from the use of this wild additive.

Worth while. Overworked as a term of vague approval and (with *not*) of disapproval. Strictly applicable only to actions: "Is it worth while to telegraph?"

His books are not worth while.	His books are not worth reading (are not worth one's while to read; do not repay reading).

The adjective *worthwhile* (one word) is acceptable but emaciated. Use a stronger word.

a worthwhile project	a promising (useful, valuable, exciting) project

Would. Commonly used to express habitual or repeated action. ("He would get up early and prepare his own breakfast before he went to work.") But when the idea of habit or repetition is expressed, in such phrases as *once a year, every day, each Sunday,* the past tense, without *would,* is usually sufficient, and, from its brevity, more emphatic.

Once a year he would visit the old mansion.	Once a year he visited the old mansion.

In narrative writing, always indicate the transition from the general to the particular—that is, from sentences that merely state a general habit to those that express the action of a specific day or period. Failure to indicate the change will cause confusion.

> Townsend would get up early and prepare his own breakfast. If the day was cold, he filled the stove and had a warm fire burning before he left the house. On his way out to the garage, he noticed that there were footprints in the new-fallen snow on the porch.

The reader is lost, having received no signal that Townsend has changed from a mere man of habit to a man who has seen a particular thing on a particular day.

> Townsend would get up early and prepare his own breakfast. If the day was cold, he filled the stove and had a warm fire burning before he left the house. One morning in January, on his way out to the garage, he noticed footprints in the new-fallen snow on the porch.

V

An Approach to Style
(With a List of Reminders)

UP TO this point, the book has been concerned with what is correct, or acceptable, in the use of English. In this final chapter, we approach style in its broader meaning: style in the sense of what is distinguished and distinguishing. Here we leave solid ground. Who can confidently say what ignites a certain combination of words, causing them to explode in the mind? Who knows why certain notes in music are capable of stirring the listener deeply, though the same notes slightly rearranged are impotent? These are high mysteries, and this chapter is a mystery story, thinly disguised. There is no satisfactory explanation of style, no infallible guide to good writing, no assurance that a person who thinks clearly will be able to write clearly, no key that unlocks the door, no inflexible rule by which writers may shape their course. Writers will often find themselves steering by stars that are disturbingly in motion.

The preceding chapters contain instructions drawn from established English usage; this one contains advice drawn from a writer's experience of writing. Since the book is a rule book, these cautionary remarks, these subtly dangerous hints, are presented in the form of rules, but they are, in essence, mere gentle reminders: they state what most of us know and at times forget.

Style is an increment in writing. When we speak of Fitzgerald's style, we don't mean his command of the relative pronoun, we mean the sound his words make on paper. All writers, by the way they use the language, reveal something of their spirits, their habits, their capacities, and their biases. This is inevitable as well as enjoyable. All writing is communication; creative writing is communication through revelation—it is the Self escaping into the open. No writer long remains incognito.

If you doubt that style is something of a mystery, try rewriting a familiar sentence and see what happens. Any much-quoted sentence will do. Suppose we take "These are the times that try men's souls." Here we have eight short, easy words, forming a simple declarative sentence. The sentence contains no flashy ingredient such as "Damn the torpedoes!" and the words, as you see, are ordinary. Yet in that arrangement, they have shown great durability; the sentence is into its third century. Now compare a few variations:

> Times like these try men's souls.
>
> How trying it is to live in these times!
>
> These are trying times for men's souls.
>
> Soulwise, these are trying times.

It seems unlikely that Thomas Paine could have made his sentiment stick if he had couched it in any of these forms. But why not? No fault of grammar can be detected in them, and in every case the meaning is clear. Each version is correct, and each, for some reason that we can't readily put our finger on, is marked for oblivion. We could, of course, talk about "rhythm" and "cadence," but the talk would be vague and unconvincing. We could declare *soulwise* to be a silly word, inappropriate to the occasion; but even that won't do—it does not answer the main question. Are we even sure *soulwise* is silly? If *otherwise* is a serviceable word, what's the matter with *soulwise?*

Here is another sentence, this one by a later Tom. It is not a famous sentence, although its author (Thomas Wolfe) is well known. "Quick are the mouths of earth, and quick the teeth that fed upon this loveliness." The sentence would

not take a prize for clarity, and rhetorically it is at the oppo-
site pole from "These are the times." Try it in a different
form, without the inversions:

> The mouths of earth are quick, and the teeth that fed
> upon this loveliness are quick, too.

The author's meaning is still intact, but not his overpower-
ing emotion. What was poetical and sensuous has become
prosy and wooden; instead of the secret sounds of beauty,
we are left with the simple crunch of mastication. (Whether
Mr. Wolfe was guilty of overwriting is, of course, another
question—one that is not pertinent here.)

With some writers, style not only reveals the spirit of the
man but reveals his identity, as surely as would his finger-
prints. Here, following, are two brief passages from the
works of two American novelists. The subject in each case
is languor. In both, the words used are ordinary, and there
is nothing eccentric about the construction.

> He did not still feel weak, he was merely luxuriating in
> that supremely gutful lassitude of convalescence in which
> time, hurry, doing, did not exist, the accumulating seconds
> and minutes and hours to which in its well state the body
> is slave both waking and sleeping, now reversed and time
> now the lip-server and mendicant to the body's pleasure
> instead of the body thrall to time's headlong course.

> Manuel drank his brandy. He felt sleepy himself. It was
> too hot to go out into the town. Besides there was nothing
> to do. He wanted to see Zurito. He would go to sleep
> while he waited.

Anyone acquainted with Faulkner and Hemingway will
have recognized them in these passages and perceived
which was which. How different are their languors!

Or take two American poets, stopping at evening. One
stops by woods, the other by laughing flesh.

> My little horse must think it queer
> To stop without a farmhouse near

Between the woods and frozen lake
The darkest evening of the year.*

I have perceived that to be with those I like is enough,
To stop in company with the rest at evening is enough,
To be surrounded by beautiful, curious, breathing,
    laughing flesh is enough . . .

Because of the characteristic styles, there is little question about identity here, and if the situations were reversed, with Whitman stopping by woods and Frost by laughing flesh (not one of his regularly scheduled stops), the reader would know who was who.

Young writers often suppose that style is a garnish for the meat of prose, a sauce by which a dull dish is made palatable. Style has no such separate entity; it is nondetachable, unfilterable. The beginner should approach style warily, realizing that it is an expression of self, and should turn resolutely away from all devices that are popularly believed to indicate style—all mannerisms, tricks, adornments. The approach to style is by way of plainness, simplicity, orderliness, sincerity.

Writing is, for most, laborious and slow. The mind travels faster than the pen; consequently, writing becomes a question of learning to make occasional wing shots, bringing down the bird of thought as it flashes by. A writer is a gunner, sometimes waiting in the blind for something to come in, sometimes roaming the countryside hoping to scare something up. Like other gunners, the writer must cultivate patience, working many covers to bring down one partridge. Here, following, are some suggestions and cautionary hints that may help the beginner find the way to a satisfactory style.

1. *Place yourself in the background.*

Write in a way that draws the reader's attention to the sense and substance of the writing, rather than to the mood and temper of the author. If the writing is solid and good, the mood and temper of the writer will eventually be revealed and not at the expense of the work. Therefore, the first piece of advice is this: to achieve style, begin by affecting none—that is, place yourself in the background. A careful and honest writer does not need to worry about style. As you become proficient in the use of language, your style will emerge, because you yourself will emerge, and when this happens you will find it increasingly easy to break through the barriers that separate you from other minds, other hearts—which is, of course, the purpose of writing, as well as its principal reward. Fortunately, the act of composition, or creation, disciplines the mind; writing is one way to go about thinking, and the practice and habit of writing not only drain the mind but supply it, too.

2. *Write in a way that comes naturally.*

Write in a way that comes easily and naturally to you, using words and phrases that come readily to hand. But do not assume that because you have acted naturally your product is without flaw.

The use of language begins with imitation. The infant imitates the sounds made by its parents; the child imitates first the spoken language, then the stuff of books. The imitative life continues long after the writer is secure in the language, for it is almost impossible to avoid imitating what one admires. Never imitate consciously, but do not worry about being an imitator; take pains instead to admire what is good. Then when you write in a way that comes naturally, you will echo the halloos that bear repeating.

3. *Work from a suitable design.*

Before beginning to compose something, gauge the nature and extent of the enterprise and work from a suitable design. (See Chapter II, Rule 12.) Design informs even the simplest structure, whether of brick and steel or of prose. You raise a pup tent from one sort of vision, a cathedral from another. This does not mean that you must sit with a blueprint always in front of you, merely that you had best anticipate what you are getting into. To compose a laundry list, you can work directly from the pile of soiled garments, ticking them off one by one. But to write a biography, you will need at least a rough scheme; you cannot plunge in blindly and start ticking off fact after fact about your subject, lest you miss the forest for the trees and there be no end to your labors.

Sometimes, of course, impulse and emotion are more compelling than design. If you are deeply troubled and are composing a letter appealing for mercy or for love, you had best not attempt to organize your emotions; the prose will have a better chance if the emotions are left in disarray—which you'll probably have to do anyway, since feelings do not usually lend themselves to rearrangement. But even the kind of writing that is essentially adventurous and impetuous will on examination be found to have a secret plan: Columbus didn't just sail, he sailed west, and the New World took shape from this simple and, we now think, sensible design.

4. *Write with nouns and verbs.*

Write with nouns and verbs, not with adjectives and adverbs. The adjective hasn't been built that can pull a weak or inaccurate noun out of a tight place. This is not to disparage adjectives and adverbs; they are indispensable parts of speech. Occasionally they surprise us with their power, as in

> Up the airy mountain,
> Down the rushy glen,
> We daren't go a-hunting
> For fear of little men . . .

The nouns *mountain* and *glen* are accurate enough, but had the mountain not become airy, the glen rushy, William Allingham might never have got off the ground with his poem. In general, however, it is nouns and verbs, not their assistants, that give good writing its toughness and color.

5. *Revise and rewrite.*

Revising is part of writing. Few writers are so expert that they can produce what they are after on the first try. Quite often you will discover, on examining the completed work, that there are serious flaws in the arrangement of the material, calling for transpositions. When this is the case, a word processor can save you time and labor as you rearrange the manuscript. You can select material on your screen and move it to a more appropriate spot, or, if you cannot find the right spot, you can move the material to the end of the manuscript until you decide whether to delete it. Some writers find that working with a printed copy of the manuscript helps them to visualize the process of change; others prefer to revise entirely on screen. Above all, do not be afraid to experiment with what you have written. Save both the original and the revised versions; you can always use the computer to restore the manuscript to its original condition, should that course seem best. Remember, it is no sign of weakness or defeat that your manuscript ends up in need of major surgery. This is a common occurrence in all writing, and among the best writers.

6. *Do not overwrite.*

Rich, ornate prose is hard to digest, generally unwholesome, and sometimes nauseating. If the sickly-sweet word, the overblown phrase are your natural form of expression, as is sometimes the case, you will have to compensate for it by a show of vigor, and by writing something as meritorious as the Song of Songs, which is Solomon's.

When writing with a computer, you must guard against wordiness. The click and flow of a word processor can be seductive, and you may find yourself adding a few unnecessary words or even a whole passage just to experience the pleasure of running your fingers over the keyboard and watching your words appear on the screen. It is always a good idea to reread your writing later and ruthlessly delete the excess.

7. *Do not overstate.*

When you overstate, readers will be instantly on guard, and everything that has preceded your overstatement as well as everything that follows it will be suspect in their minds because they have lost confidence in your judgment or your poise. Overstatement is one of the common faults. A single overstatement, wherever or however it occurs, diminishes the whole, and a single carefree superlative has the power to destroy, for readers, the object of your enthusiasm.

8. *Avoid the use of qualifiers.*

Rather, very, little, pretty—these are the leeches that infest the pond of prose, sucking the blood of words. The constant use of the adjective *little* (except to indicate size) is particularly debilitating; we should all try to do a little better, we should all be very watchful of this rule, for it is a rather important one, and we are pretty sure to violate it now and then.

9. *Do not affect a breezy manner.*

The volume of writing is enormous, these days, and much of it has a sort of windiness about it, almost as though the author were in a state of euphoria. "Spontaneous me," sang Whitman, and, in his innocence, let loose the hordes of uninspired scribblers who would one day confuse spontaneity with genius.

The breezy style is often the work of an egocentric, the person who imagines that everything that comes to mind is of general interest and that uninhibited prose creates high spirits and carries the day. Open any alumni magazine, turn to the class notes, and you are quite likely to encounter old Spontaneous Me at work—an aging collegian who writes something like this:

> Well, guys, here I am again dishing the dirt about your disorderly classmates, after pa$$ing a weekend in the Big Apple trying to catch the Columbia hoops tilt and then a cab-ride from hell through the West Side casbah. And speaking of news, howzabout tossing a few primo items this way?

This is an extreme example, but the same wind blows, at lesser velocities, across vast expanses of journalistic prose. The author in this case has managed in two sentences to commit most of the unpardonable sins: he obviously has nothing to say, he is showing off and directing the attention of the reader to himself, he is using slang with neither provocation nor ingenuity, he adopts a patronizing air by throwing in the word *primo,* he is humorless (though full of fun), dull, and empty. He has not done his work. Compare his opening remarks with the following—a plunge directly into the news:

> Clyde Crawford, who stroked the varsity shell in 1958, is swinging an oar again after a lapse of forty years. Clyde resigned last spring as executive sales manager of the Indiana Flotex Company and is now a gondolier in Venice.

This, although conventional, is compact, informative, unpretentious. The writer has dug up an item of news and presented it in a straightforward manner. What the first writer tried to accomplish by cutting rhetorical capers and by breeziness, the second writer managed to achieve by good reporting, by keeping a tight rein on his material, and by staying out of the act.

10. *Use orthodox spelling.*

In ordinary composition, use orthodox spelling. Do not write *nite* for *night, thru* for *through, pleez* for *please,* unless you plan to introduce a complete system of simplified spelling and are prepared to take the consequences.

In the original edition of *The Elements of Style,* there was a chapter on spelling. In it, the author had this to say:

> The spelling of English words is not fixed and invariable, nor does it depend on any other authority than general agreement. At the present day there is practically unanimous agreement as to the spelling of most words. . . . At any given moment, however, a relatively small number of words may be spelled in more than one way. Gradually, as a rule, one of these forms comes to be generally preferred, and the less customary form comes to look

obsolete and is discarded. From time to time new forms, mostly simplifications, are introduced by innovators, and either win their place or die of neglect.

The practical objection to unaccepted and oversimplified spellings is the disfavor with which they are received by the reader. They distract his attention and exhaust his patience. He reads the form *though* automatically, without thought of its needless complexity; he reads the abbreviation *tho* and mentally supplies the missing letters, at the cost of a fraction of his attention. The writer has defeated his own purpose.

The language manages somehow to keep pace with events. A word that has taken hold in our century is *thruway;* it was born of necessity and is apparently here to stay. In combination with *way, thru* is more serviceable than *through;* it is a high-speed word for readers who are going sixty-five. *Throughway* would be too long to fit on a road sign, too slow to serve the speeding eye. It is conceivable that because of our thruways, *through* will eventually become *thru*—after many more thousands of miles of travel.

11. *Do not explain too much.*

It is seldom advisable to tell all. Be sparing, for instance, in the use of adverbs after "he said," "she replied," and the like: "he said consolingly"; "she replied grumblingly." Let the conversation itself disclose the speaker's manner or condition. Dialogue heavily weighted with adverbs after the attributive verb is cluttery and annoying. Inexperienced writers not only overwork their adverbs but load their attributives with explanatory verbs: "he consoled," "she congratulated." They do this, apparently, in the belief that the word *said* is always in need of support, or because they have been told to do it by experts in the art of bad writing.

12. *Do not construct awkward adverbs.*

Adverbs are easy to build. Take an adjective or a participle, add -*ly,* and behold! you have an adverb. But you'd probably be better off without it. Do not write *tangledly.* The word itself is a tangle. Do not even write *tiredly.* Nobody says *tangledly* and not many people say *tiredly.* Words that are not used orally are seldom the ones to put on paper.

He climbed tiredly to bed.	He climbed wearily to bed.
The lamp cord lay tangledly beneath her chair.	The lamp cord lay in tangles beneath her chair.

Do not dress words up by adding *-ly* to them, as though putting a hat on a horse.

overly	over
muchly	much
thusly	thus

13. *Make sure the reader knows who is speaking.*

Dialogue is a total loss unless you indicate who the speaker is. In long dialogue passages containing no attributives, the reader may become lost and be compelled to go back and reread in order to puzzle the thing out. Obscurity is an imposition on the reader, to say nothing of its damage to the work.

In dialogue, make sure that your attributives do not awkwardly interrupt a spoken sentence. Place them where the break would come naturally in speech—that is, where the speaker would pause for emphasis, or take a breath. The best test for locating an attributive is to speak the sentence aloud.

"Now, my boy, we shall see," he said, "how well you have learned your lesson."	"Now, my boy," he said, "we shall see how well you have learned your lesson."
"What's more, they would never," she added, "consent to the plan."	"What's more," she added, "they would never consent to the plan."

14. *Avoid fancy words.*

Avoid the elaborate, the pretentious, the coy, and the cute. Do not be tempted by a twenty-dollar word when there is a ten-center handy, ready and able. Anglo-Saxon is a livelier tongue than Latin, so use Anglo-Saxon words. In this, as in so many matters pertaining to style, one's ear must be one's guide: *gut* is a lustier noun than *intestine,* but the two words are not interchangeable, because *gut* is often inappropriate, being too coarse for the context. Never call a stomach a tummy without good reason.

If you admire fancy words, if every sky is *beauteous,* every blonde *curvaceous,* every intelligent child *prodigious,* if you are tickled by *discombobulate,* you will have a bad time with Reminder 14. What is wrong, you ask, with *beauteous?* No one knows, for sure. There is nothing wrong, really, with any word—all are good, but some are better than others. A matter of ear, a matter of reading the books that sharpen the ear.

The line between the fancy and the plain, between the atrocious and the felicitous, is sometimes alarmingly fine. The opening phrase of the Gettysburg address is close to the line, at least by our standards today, and Mr. Lincoln, knowingly or unknowingly, was flirting with disaster when he wrote "Four score and seven years ago." The President could have got into his sentence with plain "Eighty-seven"—a saving of two words and less of a strain on the listeners' powers of multiplication. But Lincoln's ear must have told him to go ahead with four score and seven. By doing so, he achieved cadence while skirting the edge of fanciness. Suppose he had blundered over the line and written, "In the year of our Lord seventeen hundred and seventy-six." His speech would have sustained a heavy blow. Or suppose he had settled for "Eighty-seven." In that case he would have got into his introductory sentence too quickly; the timing would have been bad.

The question of ear is vital. Only the writer whose ear is reliable is in a position to use bad grammar deliberately; this writer knows for sure when a colloquialism is better than formal phrasing and is able to sustain the work at a level of good taste. So cock your ear. Years ago, students were warned not to end a sentence with a preposition; time, of course, has softened that rigid decree. Not only is the preposition acceptable at the end, sometimes it is more effective in that spot than anywhere else. "A claw hammer, not an ax, was the tool he murdered her with." This is preferable to "A claw hammer, not an ax, was the tool with which he murdered her." Why? Because it sounds more violent, more like murder. A matter of ear.

And would you write "The worst tennis player around here is I" or "The worst tennis player around here is me"? The first is good grammar, the second is good judgment—although the *me* might not do in all contexts.

The split infinitive is another trick of rhetoric in which the ear must be quicker than the handbook. Some infinitives seem to improve on being split, just as a stick of round stovewood does. "I cannot bring myself to really like the fellow." The sentence is relaxed, the meaning is clear, the violation is harmless and scarcely perceptible. Put the other way, the sentence becomes stiff, needlessly formal. A matter of ear.

There are times when the ear not only guides us through difficult situations but also saves us from minor or major embarrassments of prose. The ear, for example, must decide when to omit *that* from a sentence, when to retain it. "She knew she could do it" is preferable to "She knew that she could do it"—simpler and just as clear. But in many cases the *that* is needed. "He felt that his big nose, which was sunburned, made him look ridiculous." Omit the *that* and you have "He felt his big nose. . . ."

15. *Do not use dialect unless your ear is good.*

Do not attempt to use dialect unless you are a devoted student of the tongue you hope to reproduce. If you use dialect, be consistent. The reader will become impatient or confused upon finding two or more versions of the same word or expression. In dialect it is necessary to spell phonetically, or at least ingeniously, to capture unusual inflections. Take, for example, the word *once*. It often appears in dialect writing as *oncet,* but *oncet* looks as though it should be pronounced "onset." A better spelling would be *wunst.* But if you write it *oncet* once, write it that way throughout.

The best dialect writers, by and large, are economical of their talents; they use the minimum, not the maximum, of deviation from the norm, thus sparing their readers as well as convincing them.

16. *Be clear.*

Clarity is not the prize in writing, nor is it always the principal mark of a good style. There are occasions when obscurity serves a literary yearning, if not a literary purpose, and there are writers whose mien is more overcast than clear. But since writing is communication, clarity can only be a virtue. And although there is no substitute for merit in writing, clarity comes closest to being one. Even to a writer who is being intentionally obscure or wild of tongue we can say, "Be obscure clearly! Be wild of tongue in a way we can understand!" Even to writers of market letters, telling us (but not telling us) which securities are promising, we can say, "Be cagey plainly! Be elliptical in a straightforward fashion!"

Clarity, clarity, clarity. When you become hopelessly mired in a sentence, it is best to start fresh; do not try to fight your way through against the terrible odds of syntax. Usually what is wrong is that the construction has become too involved at some point; the sentence needs to be broken apart and replaced by two or more shorter sentences.

Muddiness is not merely a disturber of prose, it is also a destroyer of life, of hope: death on the highway caused by a badly worded road sign, heartbreak among lovers caused by a misplaced phrase in a well-intentioned letter, anguish of a traveler expecting to be met at a railroad station and not being met because of a slipshod telegram. Think of the tragedies that are rooted in ambiguity, and be clear! When you say something, make sure you have said it. The chances of your having said it are only fair.

17. *Do not inject opinion.*

Unless there is a good reason for its being there, do not inject opinion into a piece of writing. We all have opinions about almost everything, and the temptation to toss them in is great. To air one's views gratuitously, however, is to imply that the demand for them is brisk, which may not be the case, and which, in any event, may not be relevant to the discussion. Opinions scattered indiscriminately about leave the mark of egotism on a work. Similarly, to air one's views at an improper time may be in bad taste. If you have received a letter inviting you to speak at the dedication of a new cat hospital, and you hate cats, your reply, declining the invitation, does not necessarily have to cover the full range of your emotions. You must make it clear that you will not attend, but you do not have to let fly at cats. The writer of the letter asked a civil question; attack cats, then, only if you can do so with good humor, good taste, and in such a way that your answer will be courteous as well as responsive. Since you are out of sympathy with cats, you may quite properly give this as a reason for not appearing at the dedicatory ceremonies of a cat hospital. But bear in mind that your opinion of cats was not sought, only your services as a speaker. Try to keep things straight.

18. *Use figures of speech sparingly.*

The simile is a common device and a useful one, but similes coming in rapid fire, one right on top of another, are more distracting than illuminating. Readers need time to catch their breath; they can't be expected to compare everything with something else, and no relief in sight.

When you use metaphor, do not mix it up. That is, don't start by calling something a swordfish and end by calling it an hourglass.

19. *Do not take shortcuts at the cost of clarity.*

Do not use initials for the names of organizations or movements unless you are certain the initials will be readily understood. Write things out. Not everyone knows that MADD means Mothers Against Drunk Driving, and even if everyone did, there are babies being born every minute who will someday encounter the name for the first time. They deserve to see the words, not simply the initials. A good rule is to start your article by writing out names in full, and then, later, when your readers have got their bearings, to shorten them.

Many shortcuts are self-defeating; they waste the reader's time instead of conserving it. There are all sorts of rhetorical stratagems and devices that attract writers who hope to be pithy, but most of them are simply bothersome. The longest way round is usually the shortest way home, and the one truly reliable shortcut in writing is to choose words that are strong and surefooted to carry readers on their way.

20. *Avoid foreign languages.*

The writer will occasionally find it convenient or necessary to borrow from other languages. Some writers, however, from sheer exuberance or a desire to show off, sprinkle their work liberally with foreign expressions, with no regard for the reader's comfort. It is a bad habit. Write in English.

21. *Prefer the standard to the offbeat.*

Young writers will be drawn at every turn toward eccentricities in language. They will hear the beat of new vocabularies, the exciting rhythms of special segments of their society, each speaking a language of its own. All of us come under the spell of these unsettling drums; the problem for beginners is to listen to them, learn the words, feel the vibrations, and not be carried away.

Youths invariably speak to other youths in a tongue of their own devising: they renovate the language with a wild vigor, as they would a basement apartment. By the time this paragraph sees print, *psyched, nerd, ripoff, dude, geek,* and *funky* will be the words of yesteryear, and we will be fielding more recent ones that have come bouncing into our speech—some of them into our dictionary as well. A new word is always up for survival. Many do survive. Others grow stale and disappear. Most are, at least in their infancy, more appropriate to conversation than to composition.

Today, the language of advertising enjoys an enormous circulation. With its deliberate infractions of grammatical

rules and its crossbreeding of the parts of speech, it pro-
foundly influences the tongues and pens of children and
adults. Your new kitchen range is so revolutionary it *obso-
letes* all other ranges. Your counter top is beautiful because
it is *accessorized* with gold-plated faucets. Your cigarette
tastes good *like* a cigarette should. And, *like the man says,*
you will want to try one. You will also, in all probability, want
to try writing that way, using that language. You do so at your
peril, for it is the language of mutilation.

Advertisers are quite understandably interested in what
they call "attention getting." The man photographed must
have lost an eye or grown a pink beard, or he must have
three arms or be sitting wrong-end-to on a horse. This tech-
nique is proper in its place, which is the world of selling, but
the young writer had best not adopt the device of mutilation in ordinary composition, whose purpose is to engage,
not paralyze, the reader's senses. Buy the gold-plated faucets
if you will, but do not accessorize your prose. To use the language well, do not begin by hacking it to bits; accept the
whole body of it, cherish its classic form, its variety, and its
richness.

Another segment of society that has constructed a language of its own is business. People in business say that
toner cartridges are *in short supply,* that they have *updated* the next shipment of these cartridges, and that they
will *finalize* their recommendations at the next meeting of
the board. They are speaking a language familiar and dear
to them. Its portentous nouns and verbs invest ordinary
events with high adventure; executives walk among toner
cartridges, caparisoned like knights. We should tolerate
them—every person of spirit wants to ride a white horse.
The only question is whether business vocabulary is help-
ful to ordinary prose. Usually, the same ideas can be expressed less formidably, if one makes the effort. A good
many of the special words of business seem designed more
to express the user's dreams than to express a precise
meaning. Not all such words, of course, can be dismissed
summarily; indeed, no word in the language can be dis-

missed offhand by anyone who has a healthy curiosity. *Update* isn't a bad word; in the right setting it is useful. In the wrong setting, though, it is destructive, and the trouble with adopting coinages too quickly is that they will bedevil one by insinuating themselves where they do not belong. This may sound like rhetorical snobbery, or plain stuffiness; but you will discover, in the course of your work, that the setting of a word is just as restrictive as the setting of a jewel. The general rule here is to prefer the standard. *Finalize,* for instance, is not standard; it is special, and it is a peculiarly fuzzy and silly word. Does it mean "terminate," or does it mean "put into final form"? One can't be sure, really, what it means, and one gets the impression that the person using it doesn't know, either, and doesn't want to know.

The special vocabularies of the law, of the military, of government are familiar to most of us. Even the world of criticism has a modest pouch of private words (*luminous, taut*), whose only virtue is that they are exceptionally nimble and can escape from the garden of meaning over the wall. Of these critical words, Wolcott Gibbs once wrote, ". . . they are detached from the language and inflated like little balloons." The young writer should learn to spot them—words that at first glance seem freighted with delicious meaning but that soon burst in air, leaving nothing but a memory of bright sound.

The language is perpetually in flux: it is a living stream, shifting, changing, receiving new strength from a thousand tributaries, losing old forms in the backwaters of time. To suggest that a young writer not swim in the main stream of this turbulence would be foolish indeed, and such is not the intent of these cautionary remarks. The intent is to suggest that in choosing between the formal and the informal, the regular and the offbeat, the general and the special, the orthodox and the heretical, the beginner err on the side of conservatism, on the side of established usage. No idiom is taboo, no accent forbidden; there is simply a better chance of doing well if the writer holds a steady course, enters the stream of English quietly, and does not thrash about.

"But," you may ask, "what if it comes natural to me to experiment rather than conform? What if I am a pioneer, or even a genius?" Answer: then be one. But do not forget that what may seem like pioneering may be merely evasion, or laziness—the disinclination to submit to discipline. Writing good standard English is no cinch, and before you have managed it you will have encountered enough rough country to satisfy even the most adventurous spirit.

Style takes its final shape more from attitudes of mind than from principles of composition, for, as an elderly practitioner once remarked, "Writing is an act of faith, not a trick of grammar." This moral observation would have no place in a rule book were it not that style *is* the writer, and therefore what you are, rather than what you know, will at last determine your style. If you write, you must believe—in the truth and worth of the scrawl, in the ability of the reader to receive and decode the message. No one can write decently who is distrustful of the reader's intelligence, or whose attitude is patronizing.

Many references have been made in this book to "the reader," who has been much in the news. It is now necessary to warn you that your concern for the reader must be pure: you must sympathize with the reader's plight (most readers are in trouble about half the time) but never seek to know the reader's wants. Your whole duty as a writer is to please and satisfy yourself, and the true writer always plays to an audience of one. Start sniffing the air, or glancing at the Trend Machine, and you are as good as dead, although you may make a nice living.

Full of belief, sustained and elevated by the power of purpose, armed with the rules of grammar, you are ready for exposure. At this point, you may well pattern yourself on the fully exposed cow of Robert Louis Stevenson's rhyme. This friendly and commendable animal, you may recall, was "blown by all the winds that pass /And wet with all the showers." And so must you as a young writer be. In our modern idiom, we would say that you must get wet all over. Mr. Stevenson, working in a plainer style, said it with felicity, and suddenly one cow, out of so many, received the gift of immortality. Like the steadfast writer, she is at home in the wind and the rain; and, thanks to one moment of felicity, she will live on and on and on.

Afterword

Will Strunk and E. B. White were unique collaborators. Unlike Gilbert and Sullivan, or Woodward and Bernstein, they worked separately and decades apart.

We have no way of knowing whether Professor Strunk took particular notice of Elwyn Brooks White, a student of his at Cornell University in 1919. Neither teacher nor pupil could have realized that their names would be linked as they now are. Nor could they have imagined that thirty-eight years after they met, White would take this little gem of a textbook that Strunk had written for his students, polish it, expand it, and transform it into a classic.

E. B. White shared Strunk's sympathy for the reader. To Strunk's do's and don'ts he added passages about the power of words and the clear expression of thoughts and feelings. To the nuts and bolts of grammar he added a rhetorical dimension.

The editors of this edition have followed in White's footsteps, once again providing fresh examples and modernizing usage where appropriate. *The Elements of Style* is still a little book, small enough and important enough to carry in your pocket, as I carry mine. It has helped me to write better. I believe it can do the same for you.

Charles Osgood

Glossary

adjectival modifier   A word, phrase, or clause that acts as an adjective in qualifying the meaning of a noun or pronoun. *Your* country; a *turn-of-the-century* style; people *who are always late.*

adjective   A word that modifies, quantifies, or otherwise describes a noun or pronoun. *Drizzly* November; midnight *dreary; only* requirement.

adverb   A word that modifies or otherwise qualifies a verb, an adjective, or another adverb. Gestures *gracefully; exceptionally* quiet engine.

adverbial phrase   A phrase that functions as an adverb. (See *phrase.*) Landon laughs *with abandon.*

agreement   The correspondence of a verb with its subject in person and number (Karen *goes* to Cal Tech; her sisters *go* to UCLA), and of a pronoun with its antecedent in person, number, and gender (As soon as Karen finished the exam, *she* picked up *her* books and left the room).

antecedent   The noun to which a pronoun refers. A pronoun and its antecedent must agree in person, number, and gender. Michael and *his* teammates moved off campus.

appositive   A noun or noun phrase that renames or adds identifying information to a noun it immediately follows. His brother, *an accountant with Arthur Andersen,* was recently promoted.

articles   The words *a, an,* and *the,* which signal or introduce nouns. The definite article *the* refers to a particular item: *the* report. The indefinite articles *a* and *an* refer to a general item or one not already mentioned: *an* apple.

auxiliary verb   A verb that combines with the main verb to show differences in tense, person, and voice. The most common auxiliaries are forms of *be, do,* and *have.* I *am* going; we *did* not go; they *have* gone. (See also *modal auxiliaries.*)

case   The form of a noun or pronoun that reflects its grammatical function in a sentence as subject (*they*), object (*them*), or possessor (*their*). *She* gave *her* employees a raise that pleased *them* greatly.

clause   A group of related words that contains a subject and predicate. *Moths swarm* around a burning candle. While *she was taking* the test, *Karen muttered* to herself.

colloquialism   A word or expression appropriate to informal conversation but not usually suitable for academic or business writing. They wanted to *get even* (instead of they wanted to *retaliate*).

complement   A word or phrase (especially a noun or adjective) that completes the predicate. **Subject complements** complete linking verbs and rename or describe the subject: Martha is my *neighbor.* She seems *shy.* **Object complements** complete transitive verbs by describing or renaming the direct object: They found the play *exciting.* Robert considers Mary *a wonderful wife.*

compound sentence   Two or more independent clauses joined by a coordinating conjunction, a correlative conjunction, or a semicolon. *Caesar conquered Gaul,* but *Alexander the Great conquered the world.*

compound subject   Two or more simple subjects joined by a coordinating or correlative conjunction. *Hemingway and Fitzgerald* had little in common.

conjunction   A word that joins words, phrases, clauses, or sentences. The coordinating conjunctions, *and, but, or, nor, yet, so, for,* join grammatically equivalent elements. Correlative conjunctions (*both, and; either, or; neither, nor*) join the same kinds of elements.

contraction   A shortened form of a word or group of words: *can't* for cannot; *they're* for they are.

correlative expression   See *conjunction.*

dependent clause   A group of words that includes a subject and verb but is subordinate to an independent clause in a sentence. Dependent clauses begin with either a subordinating conjunction, such as *if, because, since,* or a relative pronoun, such as *who, which, that. When it gets dark,* we'll find a restaurant *that has music.*

direct object   A noun or pronoun that receives the action of a transitive verb. Pearson publishes *books.*

gerund   The *-ing* form of a verb that functions as a noun: *Hiking* is good exercise. She was praised for her *playing.*

indefinite pronoun   A pronoun that refers to an unspecified person (*anybody*) or thing (*something*).

independent clause   A group of words with a subject and verb that can stand alone as a sentence. *Raccoons steal food.*

indirect object   A noun or pronoun that indicates to whom or for whom, to what or for what the action of a transitive verb is performed. I asked *her* a question. Ed gave *the door* a kick.

infinitive/split infinitive   In the present tense, a verb phrase consisting of *to* followed by the base form of the verb (*to write*). A split infinitive occurs when one or more words separate *to* and the verb (*to boldly go*).

intransitive verb   A verb that does not take a direct object. His nerve *failed.*

linking verb   A verb that joins the subject of a sentence to its complement. Professor Chapman *is* a philosophy teacher. They *were* ecstatic.

loose sentence   A sentence that begins with the main idea and then attaches modifiers, qualifiers, and additional details: He was determined to succeed, with or without the promotion he was hoping for and in spite of the difficulties he was confronting at every turn.

main clause   An independent clause, which can stand alone as a grammatically complete sentence. Grammarians quibble.

modal auxiliaries   Any of the verbs that combine with the main verb to express necessity (*must*), obligation (*should*), permission (*may*), probability (*might*), possibility (*could*), ability (*can*), or tentativeness (*would*). Mary *might* wash the car.

modifier   A word or phrase that qualifies, describes, or limits the meaning of a word, phrase, or clause. *Frayed* ribbon, *dancing* flowers, *worldly* wisdom.

nominative pronoun   A pronoun that functions as a subject or a subject complement: *I, we, you, he, she, it, they, who.*

nonrestrictive modifier   A phrase or clause that does not limit or restrict the essential meaning of the element it modifies. My youngest niece, *who lives in Ann Arbor,* is a magazine editor.

noun   A word that names a person, place, thing, or idea. Most nouns have a plural form and a possessive form. *Carol;* the *park;* the *cup; democracy.*

number   A feature of nouns, pronouns, and a few verbs, referring to singular or plural. A subject and its corresponding verb must be consistent in number; a pronoun should agree in number with its antecedent. A solo *flute plays;* two *oboes join* in.

object The noun or pronoun that completes a prepositional phrase or the meaning of a transitive verb. (See also *direct object, indirect object,* and *preposition.*) Frost offered *his audience a poetic performance* they would likely never forget.

participial phrase A present or past participle with accompanying modifiers, objects, or complements. The buzzards, *circling with sinister determination,* squawked loudly.

participle A verbal that functions as an adjective. Present participles end in *-ing* (*brimming*); past participles typically end in *-d* or *-ed* (*injured*) or *-en* (*broken*) but may appear in other forms (*brought, been, gone*).

periodic sentence A sentence that expresses the main idea at the end. With or without their parents' consent, and whether or not they receive the assignment relocation they requested, *they are determined to get married.*

phrase A group of related words that functions as a unit but lacks a subject, a verb, or both. *Without the resources to continue.*

possessive The case of nouns and pronouns that indicates ownership or possession (*Harold's, ours, mine*).

predicate The verb and its related words in a clause or sentence. The predicate expresses what the subject does, experiences, or is. Birds *fly.* The partygoers *celebrated wildly for a long time.*

preposition A word that relates its object (a noun, pronoun, or *-ing* verb form) to another word in the sentence. She is the leader *of* our group. We opened the door *by* picking the lock. She went *out* the window.

prepositional phrase A group of words consisting of a preposition, its object, and any of the object's modifiers. Georgia *on my mind.*

principal verb　The predicating verb in a main clause or sentence.

pronominal possessive　Possessive pronouns such as *hers, its,* and *theirs.*

proper noun　The name of a particular person (*Frank Sinatra*), place (*Boston*), or thing (*Moby Dick*). Proper nouns are capitalized. Common nouns name classes of people (*singers*), places (*cities*), or things (*books*) and are not capitalized.

relative clause　A clause introduced by a relative pronoun, such as *who, which, that,* or by a relative adverb, such as *where, when, why.*

relative pronoun　A pronoun that connects a dependent clause to a main clause in a sentence: *who, whom, whose, which, that, what, whoever, whomever, whichever,* and *whatever.*

restrictive term, element, clause　A phrase or clause that limits the essential meaning of the sentence element it modifies or identifies. Professional athletes *who perform exceptionally* should earn stratospheric salaries. Since there are no commas before and after the italicized clause, the italicized clause is restrictive and suggests that only those athletes who perform exceptionally are entitled to such salaries. If commas were added before *who* and after *exceptionally,* the clause would be nonrestrictive and would suggest that *all* professional athletes should receive stratospheric salaries.

sentence fragment　A group of words that is not grammatically a complete sentence but is punctuated as one: *Because it mattered greatly.*

subject　The noun or pronoun that indicates what a sentence is about, and which the principal verb of a sentence elaborates. *The new Steven Spielberg movie* is a box office hit.

subordinate clause   A clause dependent on the main clause in a sentence. *After we finish our work,* we will go out for dinner.

syntax   The order or arrangement of words in a sentence. Syntax may exhibit parallelism (*I came, I saw, I conquered*), inversion (*Whose woods these are I think I know*), or other formal characteristics.

tense   The time of a verb's action or state of being, such as past, present, or future. *Saw, see, will see.*

transition   A word or group of words that aids coherence in writing by showing the connections between ideas. William Carlos Williams was influenced by the poetry of Walt Whitman. *Moreover,* Williams's emphasis on the present and the immediacy of the ordinary represented a rejection of the poetic stance and style of his contemporary T. S. Eliot. *In addition,* Williams's poetry . . . .

transitive verb   A verb that requires a direct object to complete its meaning: They *washed* their new car. An *intransitive verb* does not require an object to complete its meaning: The audience *laughed.* Many verbs can be both: The wind *blew* furiously. My car *blew* a gasket.

verb   A word or group of words that expresses the action or indicates the state of being of the subject. Verbs *activate* sentences.

verbal   A verb form that functions in a sentence as a noun, an adjective, or an adverb rather than as a principal verb. *Thinking* can be fun. An *embroidered* handkerchief. (See also *gerund, infinitive,* and *participle.*)

voice   The attribute of a verb that indicates whether its subject is active (Janet *played* the guitar) or passive (The guitar *was played* by Janet).

Glossary prepared by Robert DiYanni.

Index